VARIABLE AIR VOLUME MANUAL

Second Edition

VARIABLE AIR VOLUME MANUAL

Second Edition

Herb Wendes, P.E.

Published by
THE FAIRMONT PRESS, INC.
700 Indian Trail
Lilburn, GA 30247

Library of Congress Cataloging-in-Publication Data

Wendes, Herbert.
Variable air volume manual / Herb Wendes. -- 2nd ed.
p. cm.
Includes index.
ISBN 0-88173-196-X
1. Variable air volume systems (Air conditioning) I. Title.

TH7687.95.W46 1994 697--dc20 94-18941
CIP

Variable Air Volume Manual by Herb Wendes, Second Edition.

Published by The Fairmont Press, Inc.
700 Indian Trail
Lilburn, GA 30247

Printed in the United States of America

10 9 8 7 6 5 4 3 2 1

ISBN 0-88173-196-X FP

ISBN 0-13-188004-7 PH

Distributed by PTR Prentice Hall
Prentice-Hall, Inc.
A Paramount Communications Company
Englewood Cliffs, NJ 07632

Prentice-Hall International (UK) Limited, London
Prentice-Hall of Australia Pty. Limited, Sydney
Prentice-Hall Canada Inc., Toronto
Prentice-Hall Hispanoamericana, S.A., Mexico
Prentice-Hall of India Private Limited, New Delhi
Prentice-Hall of Japan, Inc., Tokyo
Simon & Schuster Asia Pte. Ltd., Singapore
Editora Prentice-Hall do Brasil, Ltda., Rio de Janeiro

CONTENTS

1. INTRODUCTION TO VAV SYSTEMS 1
   - Good News and Bad News 1
   - Problems 2
   - Basic Concept of VAV 4
   - Shifting Sun Load 5
   - Profiles of Different Cooling Load Factors 7
   - Various Total Cooling Load Profiles for Different Conditions 8
   - Zone Loads for Different Times of Day 9

2. TYPES OF VARIABLE VOLUME SYSTEMS 11
   - Understanding VAV Systems 11
   - Ductwork Pressure and Velocity Classifications 14
   - Basic Components of a VAV System 15
   - Types of VAV Systems
     - Cooling-Only Interior Systems 16
     - Combined Interior and Perimeter Systems 18
     - Separate Interior and Perimeter Systems 19
     - Dual Duct, Single Fan VAV Systems 21
     - Dual Duct, Dual Fan VAV Systems 21
     - Fan Powered Systems 23
     - VAV Induction Systems 25
     - Multizone VAV Systems 25
     - System Powered VAV Systems 27
     - "Riding The Fan Curve" Systems 29
     - Fan Discharge Damper Systems 30
     - Terminal Bypass Systems 30
     - Damper Terminal Systems 31

3. TYPES OF VARIABLE AIR VOLUME TERMINALS 33
   - Classifications of VAV Terminals 33
   - Constant Air Volume Terminals 35

VAV Terminals...36
Types of VAV Terminals 38
Cooling Only Terminal.......................................39
Reheat VAV Terminal...40
Fan Powered Terminal..41
Duct Duct VAV Terminal......................................43
VAV Induction Units..43
System Powered Boxes.......................................45
Bypass Terminals..46
Damper Terminals 47
Volume Limiting Terminals.................................47
Calibrating Pneumatic VAV Boxes 48
Calibrating Electric VAV Boxes 49
Setting Static Pressure Sensors in Pneumatic Systems.....50

4. METHODS OF VARYING FAN VOLUME..........................51
Percent Horsepower Saved 52
Minimum Motor Amp Draws..52
Riding Fan Curve with a Forward Curve
Fan and Terminal Throttling 53
Factors Affecting Operating Costs..............................55
Riding Fan Curve with a Forward Curve
Fan and Discharge Dampers................................55
Inlet Vane Dampers 56
Motor Inverters 57
Variable Pitch...59
Vaneaxial Fans...59
Variable Pitch Motor Sheaves 61
Eddy Current Drives...62
Internal Wheel Shrouds..62
Fan Basics
Centrifugal Fans 64
Types of Centrifugal Wheels...............................66
Van Axial Fans...73
Fan Performance...73
Changing Fan CFM's 78

5. TYPES OF VAV CONTROL SYSTEMS 81
Basic Types of Control Systems 81
VAV Control Sub-Systems 81
Supply Air Seasonal Temperature Requirements 83
Basic Components of VAV Control System 84
Temperature Control System Components 85
Control System Symbols 86
Pneumatic Systems 87
DDC VAV Systems 87
Energy Management Systems 88
DDC VAV Control System Components
Check-off List 92
VAV Point List 93
Return Air Fan Tracking 94
Automatic Control of Mixed and Supply Air
Temperature Resets 95
Pressure Dependent Fan Terminal Systems 96
Coil Controls 100
Morning Warm-Up and Night Heating 105
Control Costs 107
Budget Estimating Controls in New Buildings 108
Three-Way Control Valves 110

6. VARIABLE VOLUME
FUME HOOD EXHAUST SYSTEMS 111
Problems 111
Sash Sensing 113
Features of VAV Lab Control 117
Integrated Laboratory Temperature and
Pressure Control 118
Background Knowledge for VAV Exhausts 120
Air Pollution Equipment 124
Instruments to Use for Balancing
Fume Exhaust Systems 125
Balancing Procedure 126

7. VAV CONVERSIONS 129
Potential Areas of VAV Savings 129

Budget Estimating Operating Costs....132
Procedure for Evaluating a VAV Conversion....137
Energy Auditing Procedures....143
Sample Audit and Forms....150
Budget Estimating Energy Retrofit Costs....166

8. DESIGNING VAV SYSTEMS....171
Overall VAV Design Procedure....171
VAV Duct Design....173
General Requirements....173
Design Criteria....173
Air Quantities....175
Throttling Ratios....177
Diversity....177
Return Plenum Ceilings....177
Perimeter Air Systems....178
Outlet Selection....178
Calculating Seasonal Energy Consumption....180
Calculating Peak Heating and Cooling Loads....186
Weather Data....188
R Factors....189
Occupants....190
Seasonal Energy Cost Formulas....192
Heating Degree Days....194
Cooling Load Factors....199
Bin Method....201

9. TESTING AND BALANCING VAV SYSTEMS....205
Balancing VAV Systems....205
Diversity in Systems....208
Preliminaries....210
Check Heart of Systems....210
Test Reports....212
Inspect System Components....215
Take Fan Reading....217
Check VAV Terminals....220
Proportionate Balance Outlets and Duct Runs....223
Final Settings and Readings at Fan....223

Instruments 224
Ductwork Leak Testing 239

10. TROUBLE SHOOTING VAV SYSTEMS 247
Typical Problems 247
Checking Procedures 250
Spaces Too Hot or Cold 252
Too Much or Too Little Air 252
Fans 254
Imbalance 256
Drafts 256
Poor Zoning 257
Debugging VAV Systems 257

11. ESTIMATING HVAC COSTS 263
Introduction 263
General Procedure 263
Estimating Ductwork 266
Estimating Piping 272
Estimating HVAC Equipment Costs and Labor 277
Estimating Insulation 307
Estimating Temperature Controls 313
Budgeting Energy Retrofit Costs 318
Estimating Testing and Balancing 322
Labor Correction Factors 324
Existing Building Considerations 326

12. RECAP 329

13. APPENDICES
Appendix A - Abbreviations and Symbols 333
Appendix B- Association Abbreviations 335
Appendix C - Air Flow and Pressure Formulas 336
Appendix D - Changing Fan CFMs and Drives 338
Appendix E - Air Heat Transfer Formulas 340
Appendix F - Changes in State of Air Formulas 342
Appendix G - Air Density Correction Factors 344

Appendix H - Converting Velocity Pressure into
Feet Per Minute (Standard Air) 345
Appendix I - Converting Velocity Pressure into
Feet per Minute (Various Temperature) 346
Appendix J - Hydronic Formulas 347
Appendix K - Psychrometric Chart 349
Appendix L- Motor Amp Draws, Efficiencies,
Power Factors, Starter Sizes 350

INDEX 353

Chapter 1

INTRODUCTION TO VAV SYSTEMS

Variable air volume systems are the most promising and versatile type of HVAC systems available today. It's an exciting approach that knows no limit in its application and is restricted only by lack of knowledge and past habits.

Variable air volume systems, in addition to helping to solve energy problems, can save 20 to 30% in building energy costs over conventional constant air volume systems. VAV systems can also reduce first costs by using smaller equipment such as fans, pumps, boilers, chillers and less costly ductwork and piping distribution systems.

Furthermore, VAV systems can provide beautiful comfort when designed, maintained and operated properly. VAV systems provide excellent flexibility in zoning and can easily be expanded or contracted, rearranged, or partially shut down without affecting the central equipment to any degree.

In the bad news department there have been problems over the past years since the conception and inception of VAV systems, not all of which have been 100% resolved or are very easy to control

> For an example of the kinds of problems the HVAC industry has been involved with over the past decade, just imagine for a moment you have some business to transact in a new multi-story building in your city or town. Let's see what it's like as you walk through the building.
>
> You tug on the door at the entrance and it doesn't budge. You yank on it with both hands and it suddenly swings open. As you enter a gale of air rushes in with you.

The air noise from the ceiling diffuser overhead disturbs you and the jet or air blasts the toupee off your head. You scramble to set it back in place. Under another diffuser the air is stagnant and still, and further down under a third diffuser a glob of cold air sinks down and chills you.

In the first office you visit, where the heat is stifling, you take off your suit jacket and loosen your tie. In the next office, you turn your collar up and shudder. It feels as frigid as an early winter day.

The air in the building feels generally stale–as if there is no ventilation.

A sheet metal man standing on a ladder with his head thrust up into the ceiling space tells you that he sure is busy with a lot of testing and adjusting on a system that is, theoretically, self-balancing. He relates other problems to your sympathetic ears, "Not enough pressure at this box to operate it, too much at that box, and over yonder the terminal regulator isn't holding the maximum cfm."

You ride the elevator up to the mechanical room to investigate the situation there and you hear air gushing up the shaft.

In the equipment room you see cardboard covering the combustion air louver. The building engineer complains of not enough heat in the mornings upon startup. Ductwork rumbles and throbs. He worries about changing the hot, humming motor on a vaneaxial fan.

We could continue the story, but the point's been made; as versatile and powerful as VAV systems might be, they haven't always operated altogether correctly in the past.

PROBLEMS RESOLVED IN LAST DECADE

1. **Fan** volume modulation improved.
2. **DDC controls** allow for more precise controls of fans.

3. **Separate** microprocessor controls of individual **terminals.**
4. Lower **intake pressures required on VAV terminals** to operate them.
5. **Inverters** used to maximize VAV energy savings.
6. Computerized **static regain** design of **ductwork** which lessens fluctuating duct pressure and noise.
7. Proper **startup** of VAV systems.
8. Proper testing and balancing.
9. More effective locations and settings of ductwork static **pressure sensors** for fan volume control.
10. More effective settings of and actual control of VAV terminal units.

PROBLEMS NOT ALWAYS RESOLVED IN VAV SYSTEMS

1. Controlling the volume of **outside air,** building **pressures** and the economizer cycle properly.

2. Lack of **comfort**; drafts, stagnation, uncontrollable air currents, low air speeds, temperature **stratification, air current stratification** and increased relative humidity.

3. Selecting the proper amount of **minimum air flow** to maintain comfort, humidification, distribution and to makeup direct exhaust air.

4. Ceiling plenum temperatures and humidity levels.

5. The **combined effect** of higher space temperatures, low air speeds and increased relative humidity can occur at the same time in VAV systems causing discomfort at times even though the three elements fall into an acceptable comfort range.

6. Many of the VAV comfort problems occur in the summer in the cooling modes because the cooling flow rates have been reduced in recent years due to energy conserving building design, such as

reduced lighting levels, high efficiency lighting, more reflective glass, better insulation and tighter buildings.

7. **Diversity** of air volume between total of terminal units and lesser fan total not always understood and compensated for.

PEOPLE PROBLEMS

A very serious area of VAV problems is caused by a lack of knowledge, experience, skills and misconception by the people involved with VAV systems, designers, manufacturers, suppliers, contractors, balancers, building managers and facilities personnel.

Someone may not know the difference between a reheat VAV terminal and a fan powered terminal.

They may not be aware of how system powered VAV systems are controlled or how pneumatic or electronic systems operate.

They may have not concept of the difference between a pressure independent and pressure dependent terminal.

There may be a lack of understanding how fans behave under different static load conditions.

They may not be aware of diversity or how to handle it.

And the list can go on for pages but that's enough to give one an idea of how VAV systems can go awry anywhere along the line.

BASIC CONCEPT OF VARIABLE AIR VOLUME

A variable air volume system automatically varies the air flow to spaces proportionately to the current space needs at a constant temperature.

Variable air volume systems automatically adjust to changing the load factors such as solar, lighting, occupancy, internal equipment, humidity, conduction outside air loads and direct exhaust systems.

Generally, a VAV system delivers air at a constant 55°F while reducing or increasing the air quantities between preset maximum and minimum flows to satisfy the space conditions. The maximum and minimum air flow conditions are calculated to not only meet the variable load situations, but to still provide the desired comfort, health conditions and code requirements.

When heating is needed in the spaces the cool air is reduced to a preset minimum first. Then the air is heated in one of three ways, with reheat coils, recirculating warmer return air at the VAV terminals, or when an external heating system, such as a baseboard heating system, is activated.

As the various VAV terminals change their delivery volumes, a static pressure sensor in the main duct monitors the changing duct pressures and directs the fan to modulate its total air flow proportionately.

Air can also be bypassed at the fan, through a bypass duct or a fan can decrease its flow through wheel over loading. Neither of these methods is considered a true VAV system because they don't save fan energy.

Variable air volume systems operate as **constant volume** systems when all terminals are at their maximum or minimum flow settings.

SHIFTING SUN LOAD

Effect of Shifting Sun Load On Different Zones of Building

As the sun swings around from east to west during the day (Figure 1-1), this shifting sun load has a dramatic effect on the exterior area cooling required.

The east, south and west perimeter exposures of the building, zones 2, 3, and 4, are most greatly effected. The maximum loads can be double or triple the minimum loads.

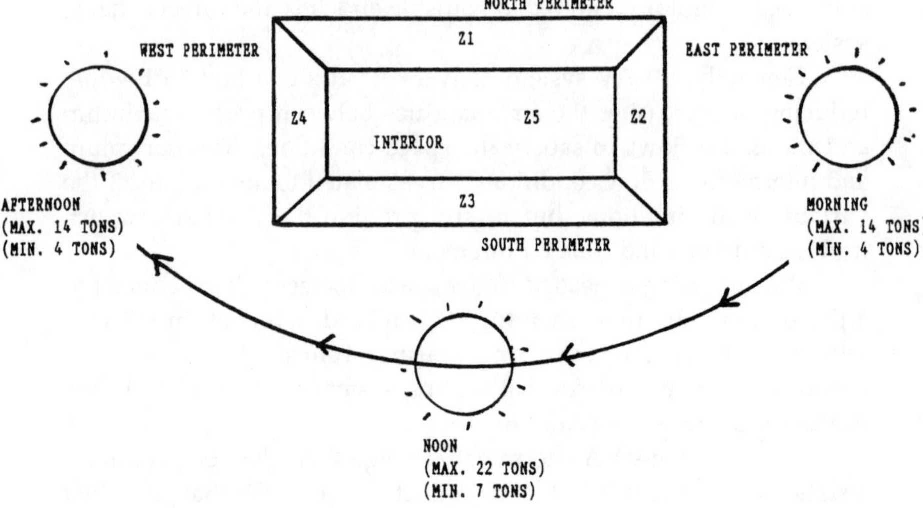

Figure 1-1. As the sun swings around from east to west during the day, this shifting sun load has a dramatic effect on the exterior area cooling required.

The interior and north perimeter zones directly under roofs are effected to some degree also by the shifting sun. The interior and north perimeter zones of in between floors in multi-story buildings are not effected by the sun's radiation.

The VAV system varies its air volume to these areas from minimum to maximum settings proportional to the loads.

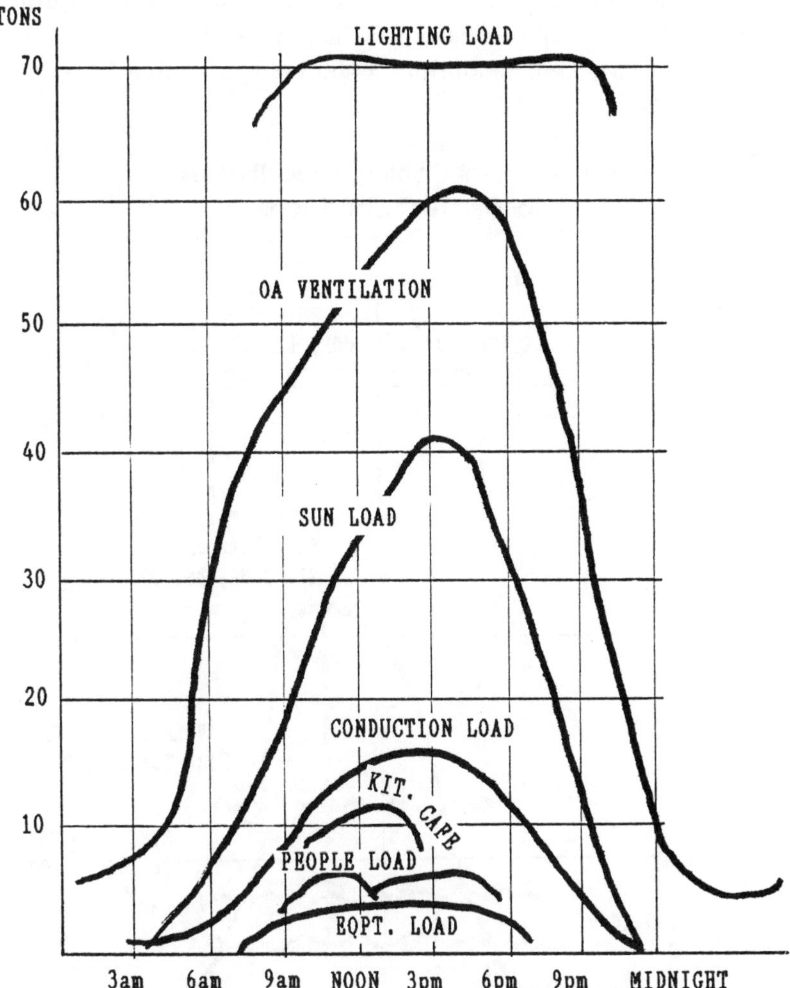

Figure 1-2. The actual total simultaneous cooling load for a building or zone varies, because the load factors do not vary proportionately with the time of day.

1. The **sun load** and the **outside air** ventilation load with its rising temperature and humidity during the day are the most volatile cooling load factors.
2. The wall, window and roof **conduction** loads are second in intensity of variation.

3. **Lighting, people** and **internal equipment** loads reach a peak at the start of the occupied day and remain rather constant during the day, then drop off sharply at the day's end.
4. Cafeteria, kitchen, laboratory **exhaust** loads have their own peak times.

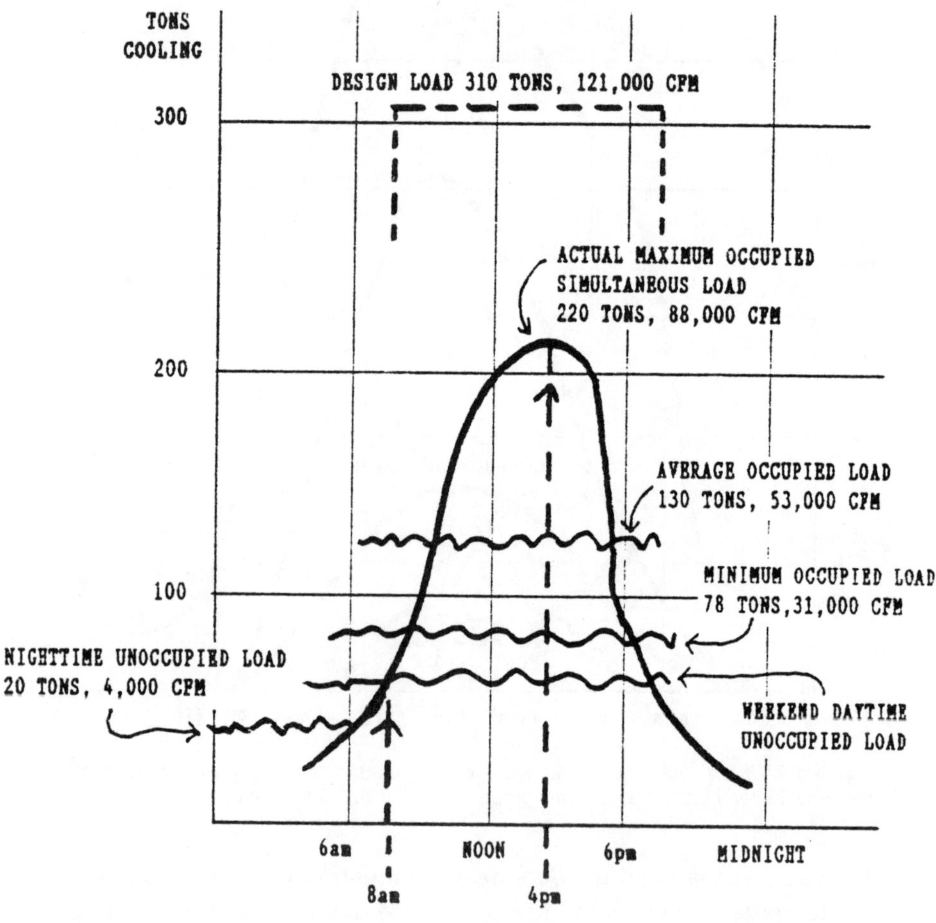

Figure 1-3. Since different areas of the building will reach their peak loads at different times during the day, the total of all the peak loads will never occur at the same time.

Also since the cooling loads from various load factors vary during the day, and since occupant conditions vary also, there are a number of different cooling load profiles as follows:

1. Actual occupied **maximum simultaneous** load, 220 tons, 88,000 cfm.
2. The actual **average** occupied load, 130 tons, 53,000 cfm.
3. The actual **minimum** occupied load, 78 tons, 31,000 cfm.
4. The night time **unoccupied** load, 20 tons, 4000 cfm.
5. The weekend daytime unoccupied load.

If the size of the cooling, heating and air distribution equipment is selected based on a lower actual maximum occupied simultaneous load, rather than on a higher design, there is a **diversity** factor.

This diversity factor can be 20 or 30% in many cases. This means that since all the areas will never peak at the same time, this sum of all the peaks will really never be needed and hence the equipment need not be sized that large.

ZONE LOADS FOR DIFFERENT TIMES OF DAY

	Zone Peaks				**Total**
	9AM	**Noon**	**3PM**	**6PM**	**Loads**
Z1 North	4	8	10 Max	8	10
Z2 East	14 Max	10	7	4	14
Z3 South	7	22 Max	20	10	22
Z4 West	4	6	14 Max	11	14
Z5 Interior	3	6	8 Max	6	8
Total Load	32 Min	52	59 Max	39	68

The above loads were calculated for four different times of the day and then summarized to determine:

1. When the maximum and minimum **simultaneous occupied loads** occur for all the areas as a lump as well as for each area individually.
2. What the total zone peak loads are.

Chapter 2
TYPES OF
VARIABLE AIR VOLUME SYSTEMS

Variable air volume systems generally are cooling-only systems used most frequently in interior areas of buildings which require cooling the year round and no heating. Variable air volume systems, however, are also used in perimeter zones and other areas requiring heat with reheat coils and recirculation capabilities.

UNDERSTANDING VAV SYSTEMS

Most HVAC systems in the past were constant air volume (CAV), variable air temperature (VAT) type systems. A residential or small commercial system is a typical example, delivering for instance, a constant 1500 CFM, while the burner or air-conditioner goes on and off changing the air temperatures to meet the cooling or heating load conditions.

A VAV system is just the opposite. It delivers air at a constant air temperature (CAT) of usually about 55°F while reducing or increasing the air quantities to satisfy the changing space loads.

The VAV system adjusts itself to changing solar, light, occupancy, internal equipment and ventilation loads by increasing or decreasing the volume of air delivered to the spaces while still maintaining health, comfort and code conditions. A predetermined maximum and minimum flow is calculated to meet the variable load.

AVERAGE VARIABLE AIR VOLUME FLOWS

VAV systems run, on an overall average, seventy percent of the peak load. Interior zones run around eighty percent of maximum. Perimeter zones vary more extensively in the cooling cycle because of the shifting solar and variable conduction loads, and run roughly an average of 60 percent of peak loads.

VARIABLE AIR VOLUME SYSTEM OPERATION PRINCIPLES

1. Basically, as stated above, VAV system varies the air flow to the spaces proportional to the heating/cooling loads at a constant 55°F.

2. When the cooling load is satisfied and possibly heating is needed in the spaces, the cool air flow is reduced to the minimum in the first stage. Then return air at a higher temperature may be recirculated. If still more heating is needed the supply air is heated with reheat coils.

3. A minimum amount of pressure is needed at the intake of pneumatic VAV terminals to operate the VAV controller to overcome the resistance of the VAV terminal and the discharge ductwork. If the pressure at the intake falls below the minimum, the terminal will not work and go to its normally 100 percent open or closed position dependent on its setting.

4. When the VAV terminals reduce their flows to the spaces, the pressure in the main supply duct increases. To reduce the pressure in the ductwork and the volume of air delivered from the fan when this occurs, some method of sensing the pressure change is needed as well as a method of fan volume control.

TRUE VARIABLE AIR VOLUME SYSTEMS VERSUS SECONDARY SYSTEMS

True VAV Systems:

A true VAV system accurately adjusts the volume of air delivered to the spaces in each zone with positive, pressure-**independent** controls on the VAV terminal. In addition it adjusts the total volume of air at the fan, also with a positive means of control, and effectively reduces fan power consumption.

Secondary VAV Systems:

Secondary VAV systems are not true VAV systems as such. The air flows are not fully controllable; they do not usually save energy and generally are used for smaller buildings in the 5- to 60-ton range.

Secondary type VAV systems control air volumes on a relatively **non-positive** basis at the terminals with pressure-**dependent**, thermally controlled or bypass terminals. Bypass ducts, bypass fans or fans with overloading wheels are methods used to control the overall flow.

MANY DIFFERENT TYPES OF VARIABLE AIR VOLUME SYSTEMS

There are many different basic types of VAV systems and variations within each basic type beyond the general separation of true and secondary categories. The following is a list of the main types:

True VAV Systems:

1. **COOLING ONLY INTERIOR VAV** SYSTEMS
2. **COMBINED INTERIOR AND PERIMETER** SYSTEMS
3. **SEPARATE INTERIOR AND PERIMETER** SYSTEMS
4. **DUAL DUCT, MIXING BOX** VAV SYSTEMS
5. **FAN POWERED** SYSTEMS
6. **VAV INDUCTION** SYSTEMS
7. **MULTIZONE** VAV SYSTEMS
8. **SYSTEM POWERED** SYSTEMS

Secondary VAV Systems:

1. RIDING THE **FAN CURVE**
2. FAN **BYPASS**
3. **TERMINAL** UNIT BYPASS
4. **DAMPER TERMINAL** SYSTEMS
5. THERMALLY CONTROLLED **VAV DAMPER TERMINALS**
6. THERMALLY CONTROLLED **SUPPLY DIFFUSERS**

DUCTWORK PRESSURE AND VELOCITY CLASSIFICATIONS

There are numerous ductwork pressure and velocities classifications in small increments spelled out by SMACNA and ASHRAE for precise ductwork construction which are more explicit than required for this book.

The general system classification based on low, medium and high pressure categories used over the years, works well as long as the basis is understood correctly.

The traditional table is as follows:

Type System	Velocity Range, fpm	Total System Static Pressure Inches of Water Gauge
Low pressure	0-2000	0"-2"
Medium Pressure	2000 up	2"-6"
High Pressure	2000 up	6" up

There are several reasons for working with total system static pressure ratings. One is that it tells the total pressure drop load on the system so that not only the correct fan can be selected in terms of performance, but that the structural strength of the fan can be compatible. A second consideration in dealing with static pressure ratings is to determine the requirement of sealing the ductwork or not. This is important because most VAV terminals won't work if there isn't sufficient pressure at the intakes of the terminals.

A common misconception is that the total system pressure is applied to the entire duct system. The total system static pressure is not the same as the static pressure in the duct system at any one point. The system static is the sum of all the resistances on the suction and discharge sides of the fan, the supply and return ductwork, outlets, dampers etc.

When low or medium pressure systems are talked about it is in the total system context. When duct pressures are talked about it is in a particular section of ductwork since the pressure is high at the suction and discharge sides of the fan and diminishes going downstream.

For example, if the total of the fan suction and discharge pressures are 4 s.p., it is considered a medium pressure VAV system. The air speed through the suction side components, however, is normally only about 500 fpm. Even though pressure drop total of the louver, dampers, coils, filter etc. might be a total of one inch, the actual duct or housing pressure is less, because of the 500 fpm air speed.

However, on the supply side, the discharge velocity of the fan might be 2500 fpm and there may be a total of three inches static pressure in the beginning section, requiring duct construction accordingly. But as the air moves downstream, the pressure diminishes. When the pressure reduces down to two inches which might be a third of the way downstream, duct construction requirements reduce accord-ingly and when the pressure is one inch, even more.

BASIC COMPONENTS OF A VAV SYSTEM

The following is a listing of typical basic components in VAV systems:

1. **VAV terminal** which varies the volume of air flowing through it based on zone heating/cooling requirements between a preset minimum and maximum. The basic parts of the terminal unit include intake, chamber, discharge sensors, controllers, dampers, linkages, and interconnecting tubing. The controller on the terminal is hooked up to a thermostat in the space of the zone being served.

2. **Fan flow modulation device** such as an inlet vane damper, discharge damper, frequency inverter on motor, etc.

3. **Main duct static pressure sensor** used to sense and measure the aggregate pressure and flow changes in the main duct due to the fluctuations in flow through the VAV terminals.

4. **The high pressure air distribution side** of a VAV system includes the main supply duct and branches to the intakes of the terminals in a pressure-independent VAV system.

5. **The low pressure air distribution side** of a VAV system includes the supply discharge ductwork from the terminal units, associated outlets and balancing dampers.

6. **Air measuring stations** can measure actual CFM of air flow through various parts of the air distribution system, such as the supply air, return air, recirculated air link, outside air duct etc. The air-measuring stations are generally constructed with a gang of pitot tubes measuring the differential velocity pressures. They may or may not have transducers on them for sending messages to fan flow controllers or to damper controllers.

7. **Return air side** of VAV system which is nearly always low pressure/low velocity ductwork. A separate return air fan is needed for larger systems.

8. **Motorized automatic dampers** for economizer cycles, outside air control, recirculation control, return air exhausted to outside, mixed air control and building pressure control.

9. **Other normal components** of HVAC system on the **suction side** of VAV system such as heating and cooling coils, filters, chambers, louvers, etc.

10. **Controllers and control panels** which can be direct digital, pneumatic or electrical.

11. **Bypass ducts** for pressure-dependent systems.

COOLING-ONLY INTERIOR VAV SYSTEMS
Single Duct

Cooling-only VAV systems throttle the air down at the terminal boxes and proportionately modulate the output at the fan. Fan flow is generally controlled with inlet vanes, frequency inverters on motors, vaneaxial fans or some other means.

Static pressure sensors or air flow measuring stations in the main duct monitor the changing pressure, and via controls, transmit a message to the fan directing it to increase or decrease its flow to meet the changing aggregate flow of the terminals.

The interior terminals modulate between minimums and maximums based on the zone cooling load.

Air flow measuring stations can be used also to monitor actual supply, return and outside air flows and adjust them accordingly to meet outside air minimums, correct tracking of return air to supply air and building or space pressure.

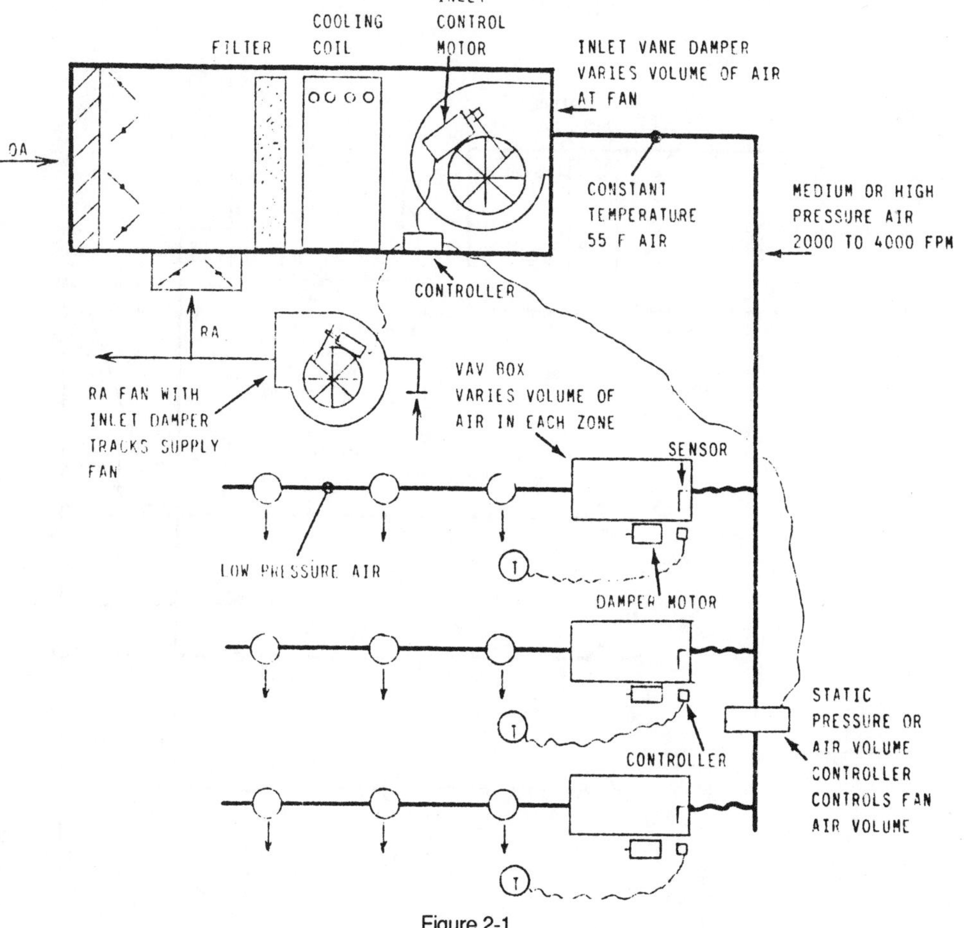

Figure 2-1

COMBINED INTERIOR AND PERIMETER SYSTEMS

Single Duct System Serving Both Cooling-Only Interior Areas and Cooling/Heating Perimeter Zones

In the combined interior and perimeter VAV system, cooling-only terminals for the interior zones are combined with reheat VAV terminals for the perimeter zones on the same system.

The interior terminals operate the same as a pure VAV cooling-only system in that they modulate between minimums and maximums based on the zone cooling load.

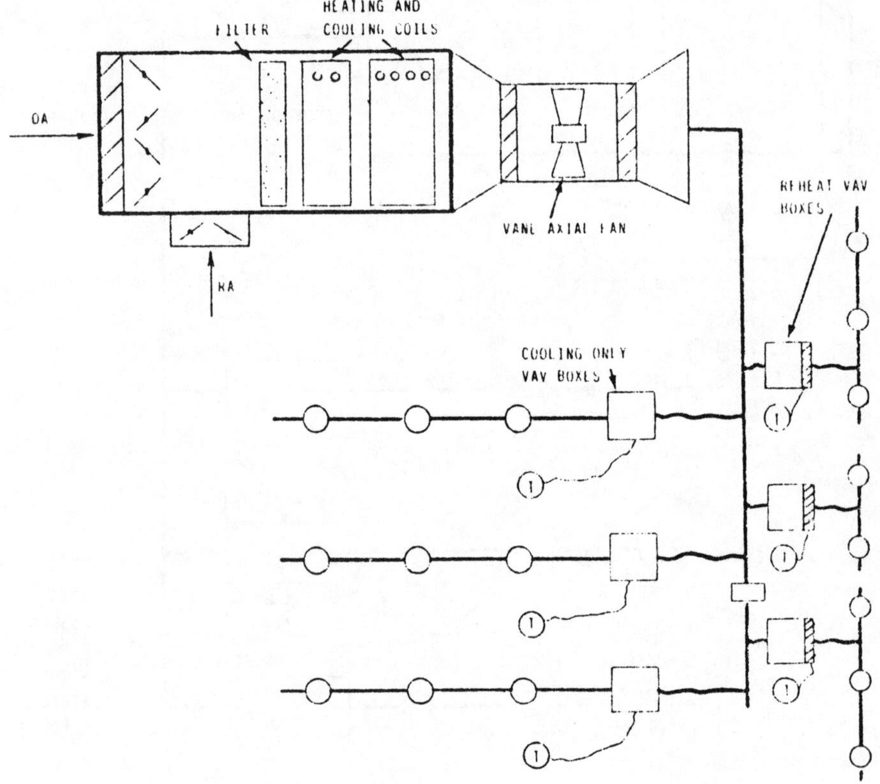

Figure 2-2

The perimeter terminals, which now have to supply heat as well as cooling because of skin heat losses, drop down to minimums first when the cooling is satisfied, and then actuate the reheat coils if more heat is needed. Reheat coils can be electric, hot water or steam.

The static pressure monitor, placed downstream in the main duct, monitors the changing supply duct pressure and instructs the fan to modulate according to whatever the total volume of air the terminals are actually delivering at any point in time.

SEPARATE INTERIOR AND PERIMETER SYSTEMS

Separate Interior Cooling-Only VAV System Combined with Independent Hydronic or Air Perimeter Systems

A very common approach in office buildings is to have a interior cooling-only VAV system combined with an independent hydronic system for heating the perimeter or with a heating-only, or heating or cooling air system for the perimeter. (See Fig. 2-3)

Perimeter hydronic systems can be baseboard, cabinet heaters or radiators. When hydronic systems are combined with cooling-only VAV interior systems, the interior cooling system normally serves the perimeter also for cooling purposes. When the cooling in exterior zone is satisfied the VAV terminals throttle down to their minimum flows first. Then if heating is needed the perimeter hydronic system is activated.

Perimeter air systems might be constant volume reheat or induction systems or even separate variable air volume reheat systems. Induction units generally have heating coils in them. If there is no cooling in the perimeter system the operation is similar to that of the hydronic approach in that the interior VAV cooling-only system serves both the interior and perimeter for cooling and goes to minimum in the perimeter zones before perimeter air heating systems are actuated.

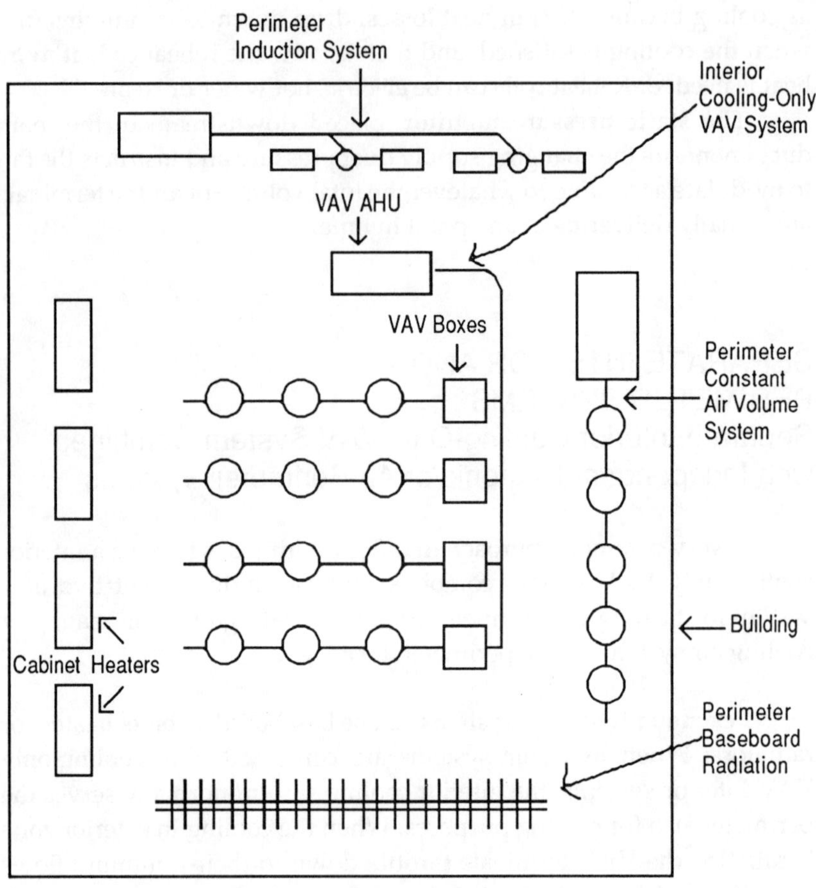

Figure 2-3. Separate interior and perimeter systems.

DUAL DUCT, SINGLE FAN VAV SYSTEM

The central air supply unit in the dual duct, single fan VAV system has two supply decks, a hot deck on top with a heating coil and a larger cold deck below with a cooling coil. A single supply fan blows the supply air through both coils and the total flow is controlled with a fan flow modulation device.

Hot and cold duct branch-offs from the main ducts feed into the hot and cold intakes of the VAV terminals.

The VAV mixing boxes modulate both the volume and temperature of air according to the space load. The cool air flow is throttled down in the terminal as the load decreases until it reaches the preset minimum. Continued decreases in the load simultaneously open up the heating damper in the terminal extracting air from the heating duct, and completely closes the cold damper. The system uses return air at first in the heating duct, and if additional heat is needed, the hot deck heating coil turns on.

A static pressure sensor is installed about two thirds the way down the main duct the same way as with other true VAV systems to sense duct pressure changes and regulate the flow at the fan. (See Figure 2-4.)

DUAL DUCT, DUAL FAN VAV SYSTEMS

1. The double duct, double fan VAV system can be one of the most energy efficient HVAC systems available. Each fan has its own independent static pressure controller and the total flow of both fans never exceeds the return air capacity.
2. In the dual duct, dual fan type VAV system there are two supply fans, one for the hot deck and the other for the cold deck.
3. The cold deck fan supplies air at a constant temperature cooled by the cooling coil or by outside air during economizer cycles.
4. The hot deck fan supplies warm return air back to the spaces. Additional heat can be added from a hot deck coil when the heating load cannot be met with the return air.
5. Only the hot deck fan needs to operate during warm up and night cycle periods.

6. Both the hot and cold deck fan share a common return air fan and duct system.
7. The VAV mixing terminals modulate the hot and cold air to meet zone loads the same way as dual duct, single fan systems. (See Figure 2-5.)

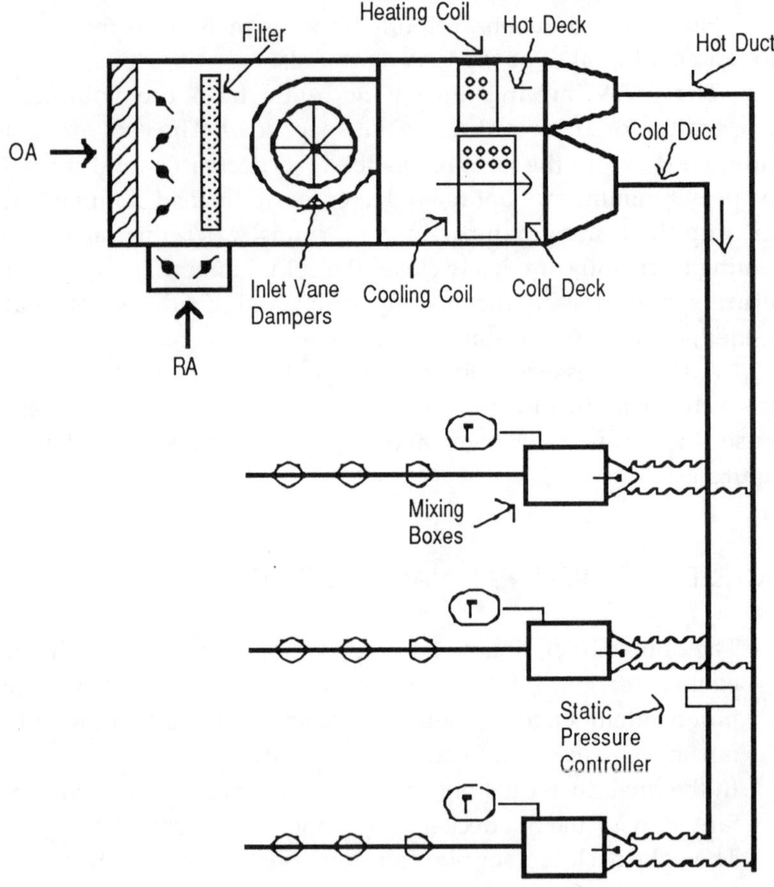

Figure 2-4. Dual Duct, Single Fan VAV System.

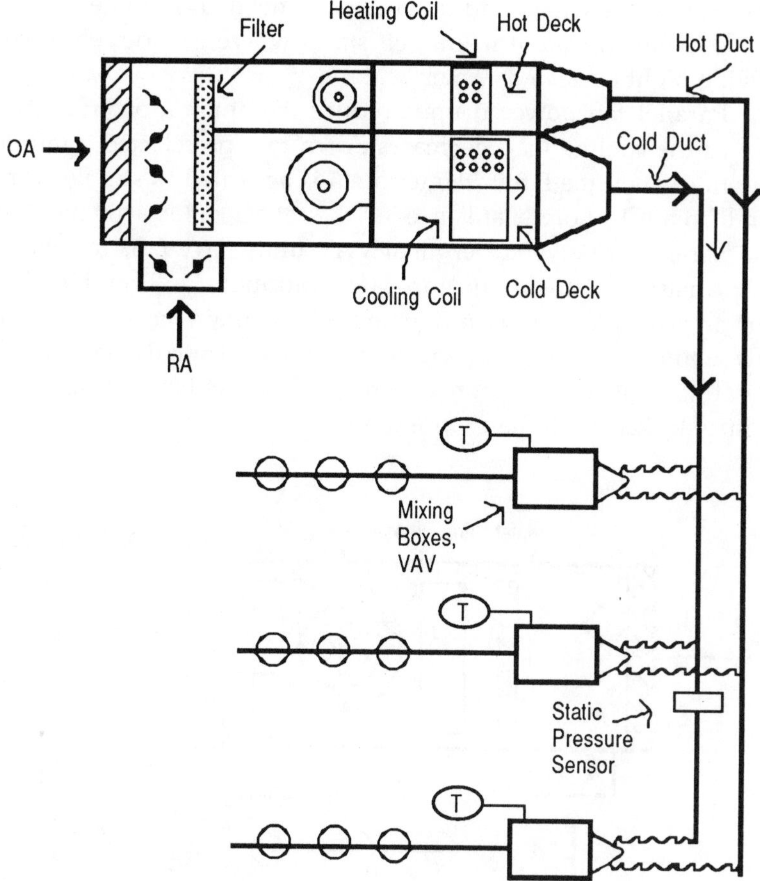

Figure 2-5. Single Duct Dual VAV System

FAN POWERED SYSTEMS

Parallel fan powered terminal VAV units have two separate parallel chambers, one for the primary cool air flow and the other for ceiling return air flow. There is a fan in the return air chamber powering the return air flow. There may or may not be a reheat coil at the discharge of the terminal.

Fan powered VAV terminals (Fig. 2-6) are mainly used in the perimeter areas where additional heat is needed in winter and are very popular in southern sun belt states where electric reheat coils provide night cycle and winter heat.

Parallel fan powered units reduce the amount of primary air flow as the cooling load decreases down to a preselected minimum flow rate. Then the fan in the terminal is activated to supply warmer return air. If more heat is needed the reheat coil is activated.

Series fan powered terminal VAV units have fans installed in the primary air stream and operate continuously. They blend the cool primary air and warm plenum air maintaining a constant air flow while adjusting temperatures. Series terminals have higher operating costs and are more expensive than parallel units and must be inter-locked with the central fan.

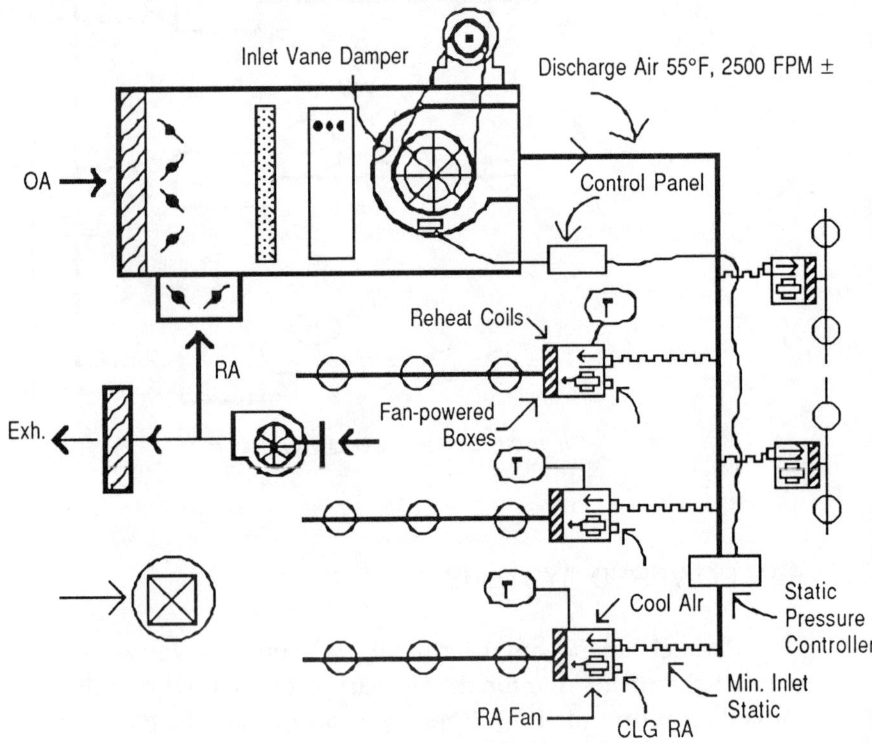

Figure 2-6. Fan Powered System.

VAV INDUCTION SYSTEMS

One good application of VAV induction systems is for hospital patient rooms.

At maximum flow the VAV induction system generally combines about one third cool primary air at a constant 55°F, with two thirds induced room return air. As the cooling load drops in the spaces it simultaneously reduces primary air and increases induced room or ceiling plenum air to maintain a constant room supply. If more heating is needed the primary cool supply air is reduced to a minimum and a reheat coil in the VAV terminal is activated.

If VAV induction units are used without reheat coils, a separate perimeter heating system is needed.

Some of the advantages of VAV induction systems are that there is slower air flow with less drafts and possible contaminated room return air need not be circulated through the system. (Fig. 2-7.)

MULTIZONE VAV SYSTEMS

Multizone VAV systems have a single central air handler with a hot and cold deck which has a series of automatic hot and cold deck dampers in the discharge for usually five to ten zone ducts. Each zone duct is a separate main duct and usually has to cover a number of rooms under one zone thermostat.

Different zone load requirements are met by mixing hot and cold air through the zone dampers at the central air handler. It is similar in operation to the dual duct system in that it combines and modulates the hot and cold air in each zone, but does it at the fan rather than at each individual mixing box.

To vary the air volume in the zones VAV terminals are used. The volume of cool air is reduced to a minimum first, then if the load decreases more, the cold and the hot deck dampers modulate temperatures. If more heat is needed the cold deck damper closes all the way and the hot deck damper goes to full open.

If the VAV terminal is in a perimeter zone, a reheat coil is activated if more heat is needed. The VAV terminals in cooling-only interiors operate the same as in a standard cooling-only system.

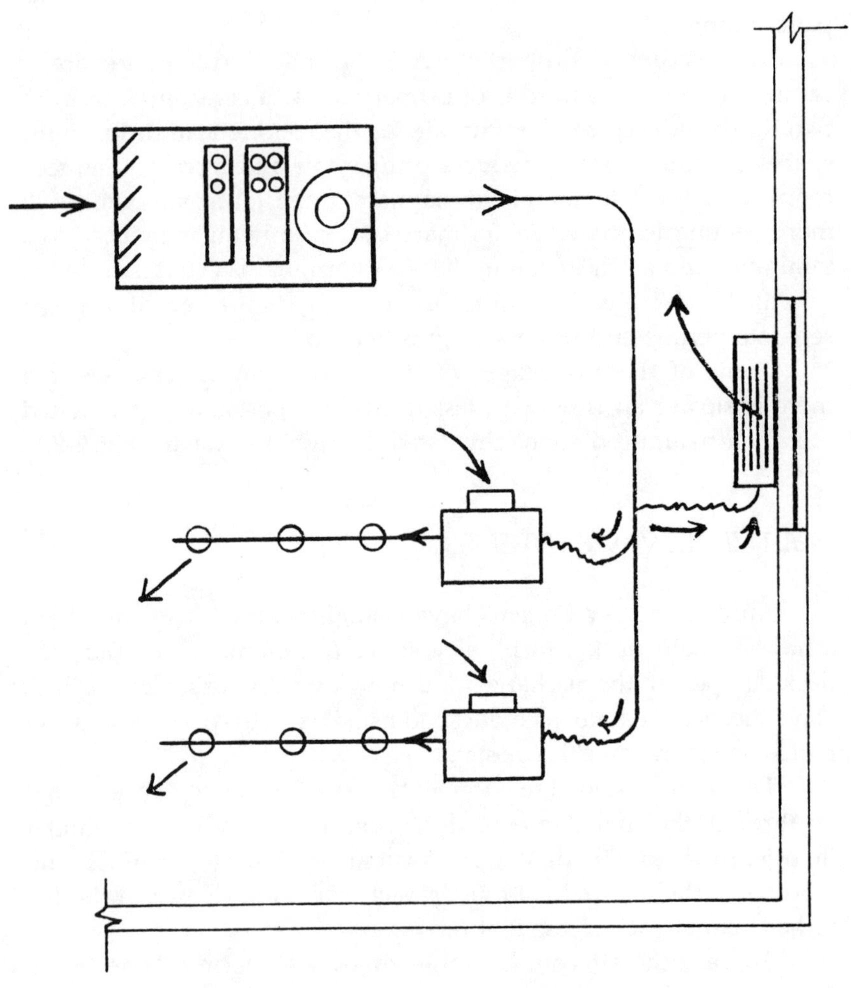

Figure 2-7. VAV Induction System.

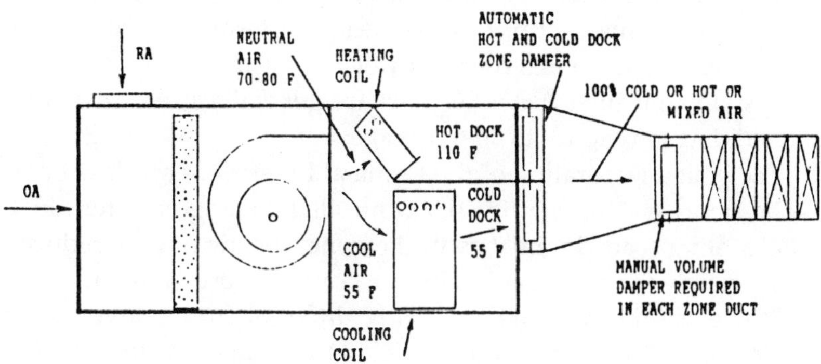

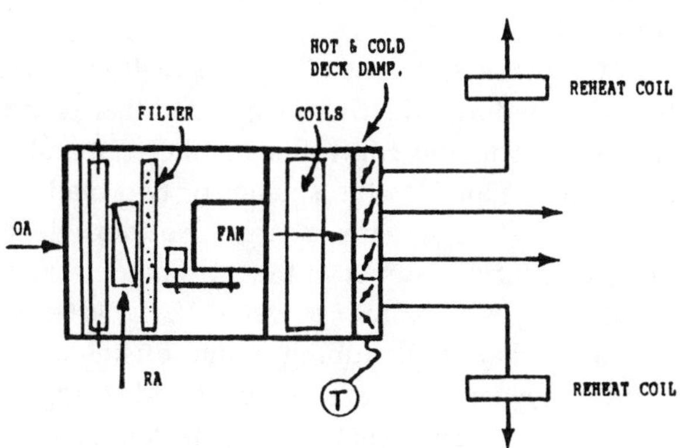

Figure 2-8. Multizone VAV System.

SYSTEM POWERED VAV SYSTEMS

A system powered VAV system uses supply duct air to power the controls on the terminal unit instead of pneumatically or electrically powered control systems. There are no separate control systems for the terminals, rather each thermostat and terminal is a separate little independent system.

To power the thermostat and the terminal unit, control tubing is tapped into the main supply duct near the terminal and system supply air is drawn out for control power.

System controlled VAV systems are generally used in smaller office buildings in the 10- to 50-ton range.

The general operation of the terminal in terms of air flow operates the same as a conventionally controlled cooling-only terminal. As the cooling load diminishes in the zone, the air flow is reduced proportionately from a preset maximum down to a preset minimum.

However, the controls are vastly different from conventional controls. The tubing is either 5/8, 3/4 or 1 inch in diameter, varying with the distance between the supply duct and the terminal. About an inch of static pressure is needed to operate the terminal and possibly 1/4 of an inch more for the discharge duct. This means the pressure in the ductwork can never fall below 1-1/4 inches W.G. or the box will fail.

In the process of controlling the air flow, "system air" from the main duct is fed into the thermostat. Another tube is connected from the thermostat to the controller on the terminal. The controller in turn meters the air into the smaller of two bladders. As the smaller bladder expands it actuates a larger bladder which is hooked up to the volume control damper in the terminal.

Advantages of system powered VAV systems are that the owner can easily change the tubing without using a steam fitter and initial control system costs are lower. **Disadvantages** are that fibers in the supply air stream can clog the stats, and reaction time is slower.

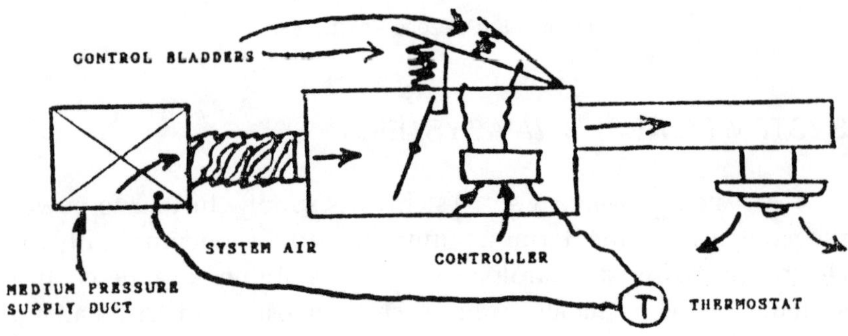

Figure 2-9. System Powered VAV Systems.

"RIDING THE FAN CURVE" VAV SYSTEMS

When the centrifugal forward curved fan wheel is overloaded with increasing system resistance, the amount of air flow is reduced without overloading the motor, within certain ranges on the fan curve. Hence, no external fan modulation is needed.

In riding the fan curve systems, pressure independent or dependent VAV terminals can be used to modulate the flows in the zones. No static pressure sensor is needed in the main duct because the fan flow reduces automatically in accordance with the duct pressure loads.

This type of system is used in smaller commercial building and systems in the 20- to 60-ton range and frequently with roof-top units.

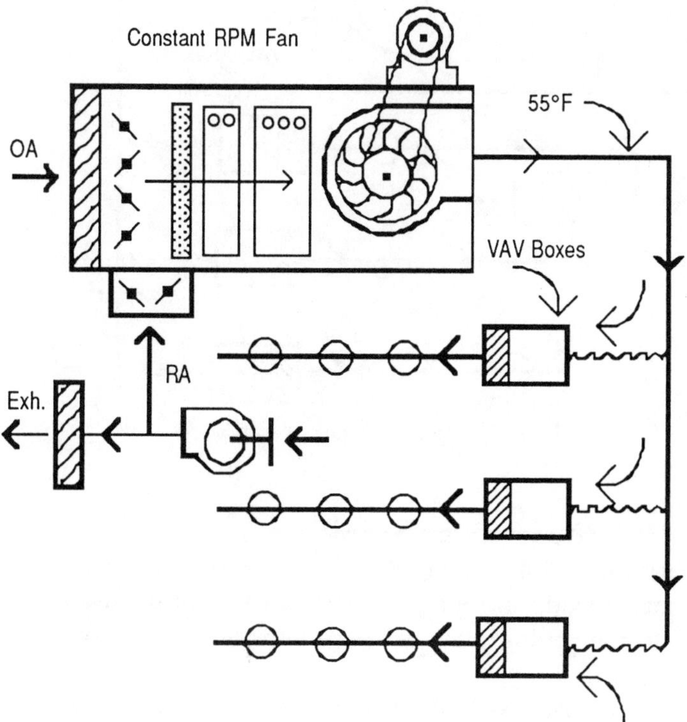

Figure 2-10. "Riding the fan curve" VAV System.

FAN DISCHARGE DAMPER SYSTEMS

In fan discharge damper systems the discharge dampers are used to modulate flow at the fan. The VAV terminals can be pressure independent or dependent. A static pressure sensor in the main duct controls the volume of air at the fan. Supply air is discharged at a constant 55°F.

Fan discharge dampers are used in smaller commercial buildings and frequently on roof-top units.

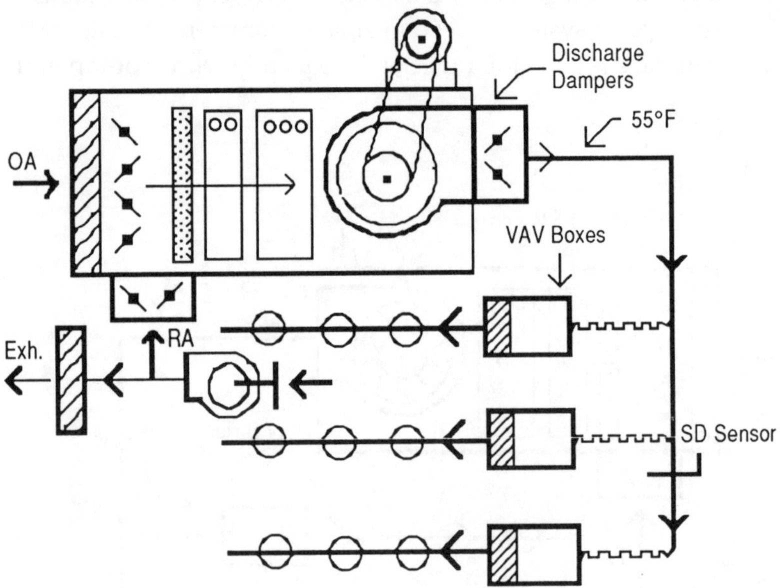

Figure 2-11. Fan Discharge System.

TERMINAL BYPASS VAV SYSTEMS

To vary the volume of air in the terminal bypass VAV system to meet reduced loads, air bypasses the discharge of the terminal, is dumped into the ceiling plenum space and then cycled back to a constant volume fan.

The terminal bypass VAV system is used for smaller systems and buildings in the 20- to 50-ton range. It provides zone control and lower initial costs but there are no fan energy savings.

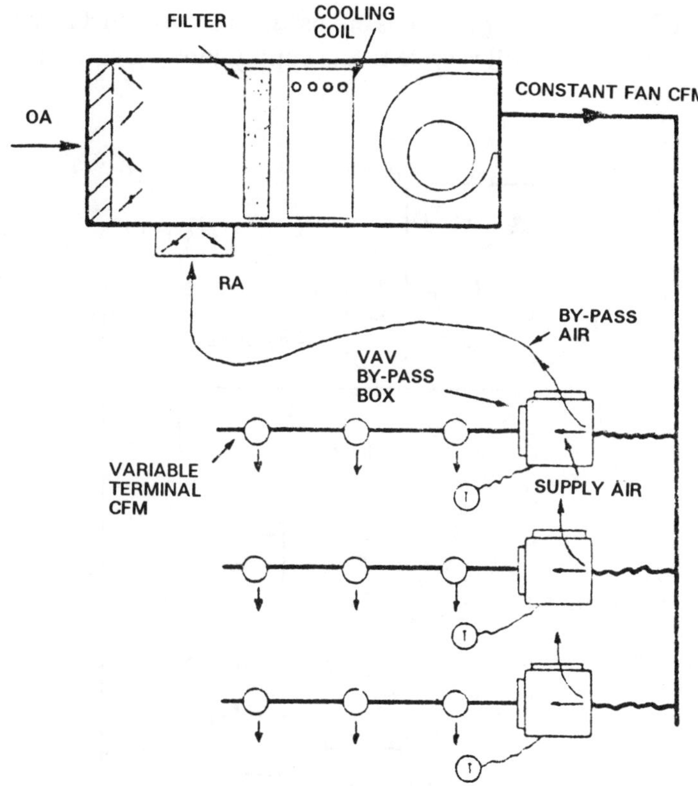

Figure 2-12. Terminal Bypass VAV System.

VAV DAMPER TERMINAL SYSTEMS

The VAV Damper Terminal system is a low pressure variable air volume system with thermally controlled, pressure dependent VAV zone dampers terminals, constant volume fans and return air bypasses. Space comfort is maintained by varying the flow of the cold or hot constant temperature air to the spaces via a zone thermostat system.

Systems are generally limited to a maximum 40 to 50 tons of cooling.

When the terminals reduce their air flow, a bypass duct or ceiling return plenum is used to maintain a constant volume at the supply fan and proper static pressure in the main supply duct.

If ceiling return air plenums are used, no bypass duct is required and the bypass air will recycle to the supply fan through the return air ceilings.

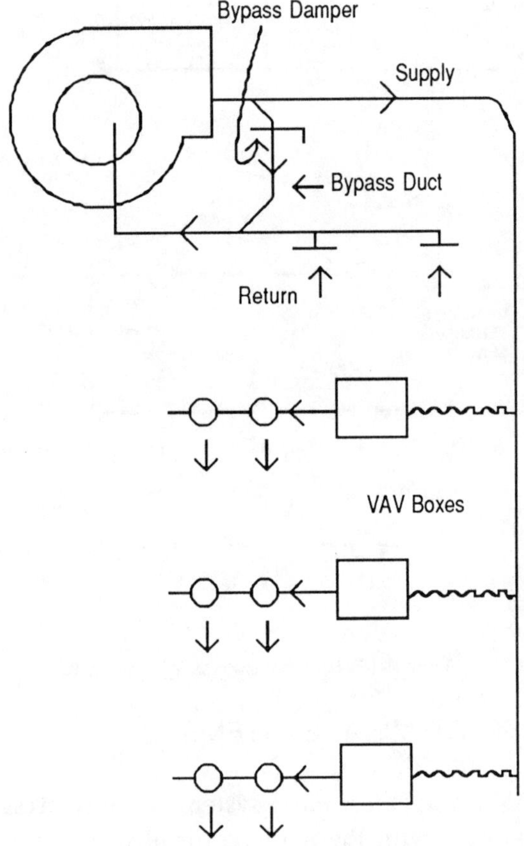

Figure 2-13. VAV Damper Terminal System.

Chapter 3
TYPES OF VARIABLE AIR VOLUME TERMINALS

The main purpose of the older constant air volume terminal is to maintain a constant volume of air while the air temperature is adjusted to meet the variations in loads. The purpose of newer VAV terminals is to adjust the air flow instead, in order to meet the variations in space load conditions, while maintaining a constant air temperature.

However, at maximum and minimum flows the VAV terminal becomes, in reality, a constant volume terminal instead of variable volume unit, because it maintains a fixed flow at these limits. Consequently the VAV terminal functions both as a VAV and CAV terminal.

There are many basic different types of VAV terminals and variations thereof such as cooling-only, reheat, fan powered, mixing boxes, single duct, double duct, induction, bypass and so on.

CLASSIFICATIONS

Variable air volume terminals are classified as to how their controls are powered, whether they are pressure independent or dependent, are they normally open or closed and are they direct or indirect acting.

Types of Control Power

Variable air volume terminals, valves, control dampers etc. can be controlled with a number of different types of control power, such as electric, pneumatic, direct digital or system power.

Pressure Independent or Dependent

The amount of air flow in pressure independent boxes is not affected by changes in static pressures at the intake.

For example if 800 CFM is to flow through a box on maximum and 200 CFM on minimum, those amounts of air will flow through whether the intake pressure is .50 inches of 6 inches at the intake.

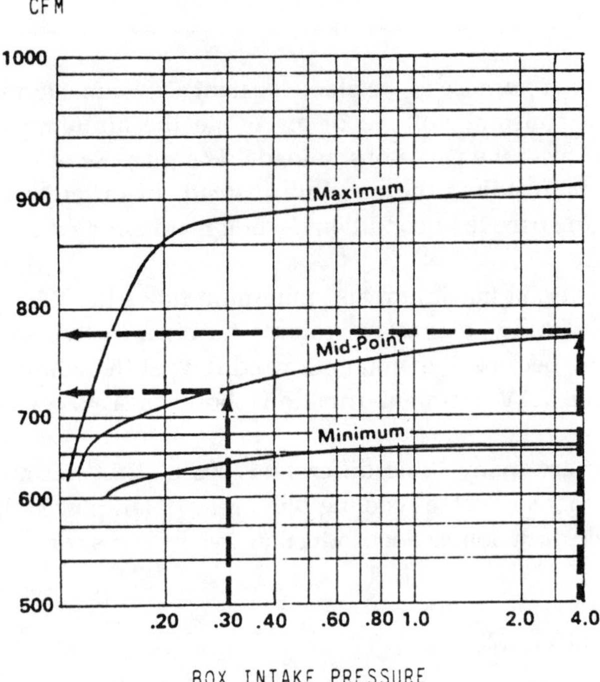

Figure 3-1. Typical pressure independent volume regulation. The variation in pressure between .3 inches and 4.0 inches is only about 5 percent.

In pressure dependent boxes the amount of air flow is dependent on the intake static pressure and varies accordingly. At 3 inches intake pressure 800 CFM might flow through the box, whereas at .5 inches only 300 CFM might flow through it.

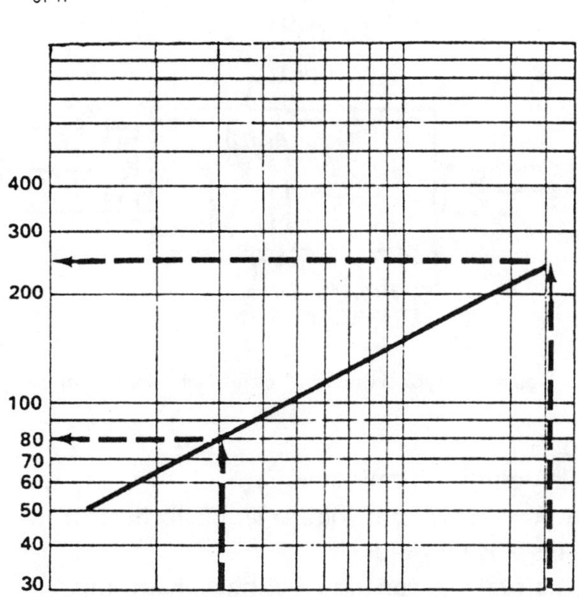

Figure 3-2. Typical pressure dependent volume regulation. At .03 inches 80 CFM flows through box and at .3 inches 250 CFM.

Normally Open or Closed

Another important aspect of VAV boxes, is whether a box is normally open or closed. The normally closed box is closed when the system is at rest and the normally open box is open when the system is at rest. A normally open box requires a reverse controller.

CONSTANT AIR VOLUME TERMINAL DEFINITION

Components

The older constant air volume terminal consists of an intake collar, housing, acoustical liner and baffling, constant volume regulator, constant volume controller, a discharge collar and a reheat coil (or no reheat coil).

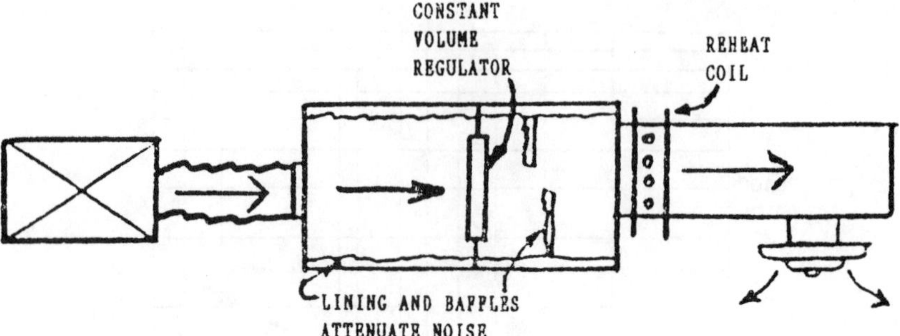

Figure 3-3. Typical Constant Volume Reheat Terminal.

Functions of Constant Air Volume Terminals

The purpose of constant volume terminals in older high pressure HVAC systems was fourfold:

- To maintain a constant volume of air flow through the terminal with a constant volume regulator at a constant 55°F delivery temperature.
- To reduce the speed and pressure of the high velocity pressure air down to low velocity and pressure levels.
- To attenuate the air noise of the high velocity air.
- To control the temperature of the air if the cooling load was satisfied by reheating the constant 55°F air or mixing it with heating primary air.

PURPOSE OF VARIABLE AIR VOLUME TERMINALS

The functions of a typical variable air volume terminal are:

- To vary the volume of air flow commensurate with the load at a constant 55°F + or – delivery temperature.
- To mix cool supply air with room or ceiling induced air to meet space loads such as with fan powered or induction units.
- To mix cold and hot primary air to meet varying space loads as with mixing boxes.

- To reduce the velocity and pressure of the air in medium pressure systems.
- To attenuate the air noise of medium or high velocity air.
- To control the temperature of the air when the cooling load is satisfied by reducing cool primary air, recirculating warmer return air, reheating or mixing primary heating air in.

COMPONENTS OF A TYPICAL VAV TERMINAL

1. Intake collar
2. Housing
3. Differential sensor in intake
4. External volume controller
5. Volume control dampers for pressure independent terminals
6. Pressure sensitive air flow throttling device for pressure dependent terminal
7. Damper motor and linkages
8. Sound attenuation lines and baffles
9. Rectangular discharge collar

TYPES OF DIFFERENTIAL SENSORS

Differential sensors located in the intake of the VAV terminal measure the total and static pressures of the incoming primary air at that point. Via the tubing hookup, the static pressure is physically subtracted from the total pressure resulting in the velocity pressure, which is generally referred to as the **differential pressure**.

This differential pressure is calibrated at the terminal manufacturer's laboratory proportional to a range of air flow for that particular type and size terminal. (See Figure 3-4.)

TYPES OF VOLUME CONTROLLERS

The volume controller located on the outside of the terminal is the intelligence device which:

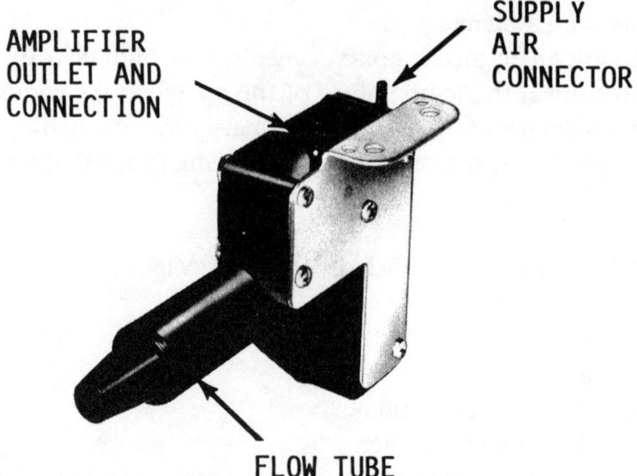

Figure 3-4. Differential Sensor

1. Reads the different intake pressures to track flow control.
2. Takes in main control air in pneumatic systems for powering controls.
3. Reads the thermostat tracking of space temperatures.
4. Meters pneumatic control air to modulate the volume control damper.
5. Allows user to set maximum and minimum flows, reads settings and limits flows accordingly.
6. Allows user to set terminal at normally open or closed position when HVAC is turned off and unit is at rest. (See Figure 3-5.)

TYPES OF VAV TERMINALS

1. **Cooling-only**, single duct
2. **Reheat** VAV terminals, single duct
3. **Fan Powered**
   a. Parallel
   b. Series
4. **Dual Duct**, mixing box
5. **Induction**, single duct

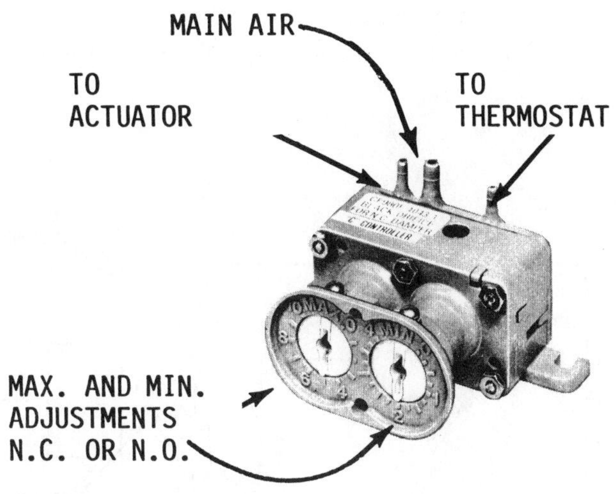

Figure 3-5. Typical Volume Controller

6. **System Powered,** single duct
7. **Bypass Terminals,** (cooling-only, single duct)
8. **Damper** section only terminal
9. **Air Valves**
10. **Volume Limiting** Terminal

Cooling-Only VAV Terminal

The cooling-only VAV terminal is generally used in interior areas of multi-story buildings needing cooling the year round and no heating.

They are single duct terminals, have all the components as previously stated and are generally pressure independent.

They provide cooling only and don't have the ability to supply warm air except in the reverse warm-up cycle. They modulate the air flow between maximum and minimum settings, reduce velocity and pressure, attenuate sound and when there is a low cooling required, reduce to a minimum flow, always at a constant supply temperature of about 55°F.

The intake sensor monitors the differential pressure and the VAV controller modulates the volume damper.

See Figure 3-6, Components of a Typical VAV Terminal.

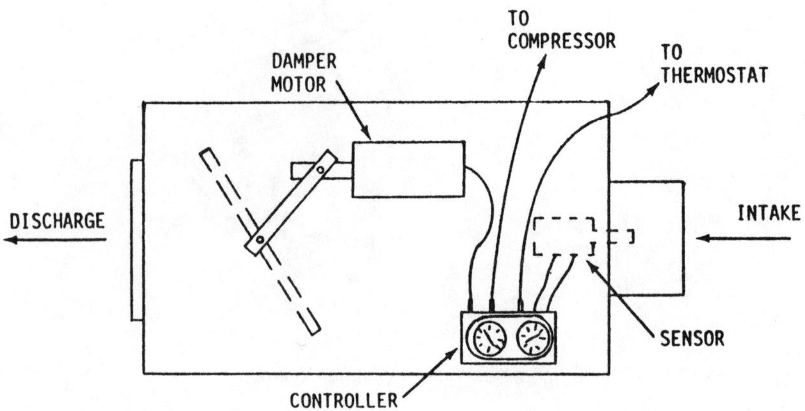

Figure 3-6. Components of a Typical VAV Terminal.

Reheat VAV Terminals

The reheat VAV terminal is generally used in exterior and under roof areas of buildings which require heating as well as cooling.

They are single duct terminals and have all the components of the standard cooling VAV terminal, plus a reheat coil at the discharge. The reheat coil can be hot water, steam or electrical. Reheat VAV terminals are generally pressure independent terminals.

They function basically as the cooling terminal, modulate the air flow between maximum and minimum settings, reduce velocity and pressure, and attenuate sound. In addition to those functions, after they reduce to minimum flow, if heat is still needed, they activate the reheat coil and heat the cool primary air.

The intake sensor monitors the intake flow into the terminal and the VAV controller modulates the volume damper.

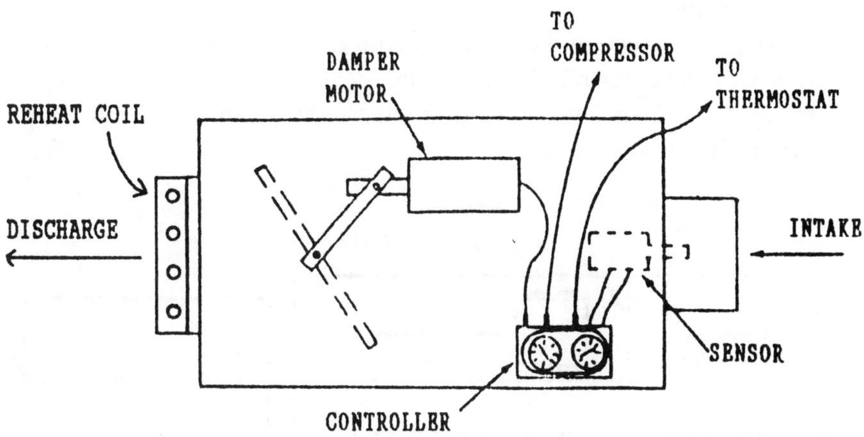

Figure 3-7. Reheat VAV Terminal

Fan Powered VAV Terminals

Fan powered VAV terminals are normally installed in plenum ceilings and used in spaces requiring heating as well as cooling, such as building exterior zones, areas under roof and in constant fan CFM areas. They are selected frequently due to their ability to produce higher room air motion during low cooling loads and during heating, compared to cooling only, VAV reheat or perimeter hydronic systems.

Components

A small terminal fan is designed into the terminal by the manufacturer to draw in warm ceiling air. System supply air flow is modulated between maximum and minimum limits to meet loads, so it is a true VAV terminal.

Reheat coils may be added to the discharge if more heat is required. Some manufacturers do not have the reheat coil option and a separate duct reheat coil is installed downstream if it is required.

The terminal fan handles not only the recirculation of plenum return air when required, but can also handle the static pressure drop of the terminal and downstream ductwork and diffusers.

Heating of exterior areas can be handled during off hours without the need to operate the primary air system fan.

Parallel Units

There are two types of fan powered VAV terminals, parallel and series. The terminal fan in the parallel unit is outside the primary air stream. The series unit has the terminal fan installed in series in the air stream.

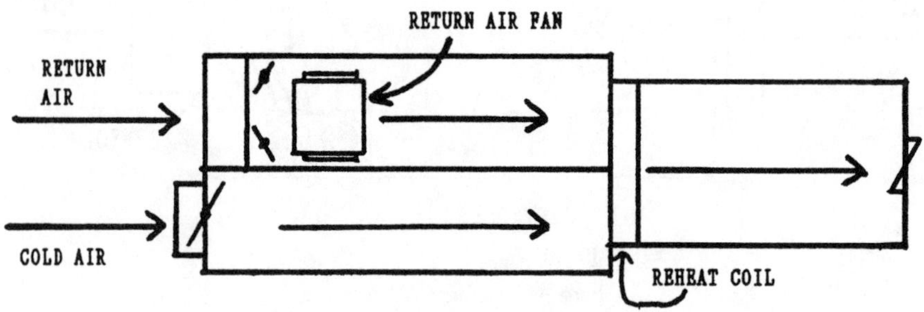

Figure 3-8. Fan Powered VAV Terminals

The parallel type is generally preferred since its fan motor runs intermittently and consumes less energy than the constantly running series terminal fan.

The sequence of operation of the parallel unit is as follows: as with cooling-only VAV terminal the air flow is reduced to a preset minimum first when the cooling load is satisfied and then the terminal fan is actuated drawing in the plenum return air and blending it with the minimum primary cool air stream. If more heat is needed the reheat coil is actuated. As the heating load is satisfied, the reverse process occurs, the reheat coil goes off, then the fan is turned off and finally the flow is increased from minimum upward as needed.

Series Units

The operation in a series fan powered terminal is similar in that primary air flow is reduced first when the cooling load is met. However, the terminal fan runs constantly and isn't turned on and off. If more heat is needed the reheat coil is then actuated.

Series units are used where a more constant CFM is required, regardless of the load, such as washrooms, hallways, entrance ways, conference rooms, etc.

Energy Consumption

Energy used by terminal fans depends upon operating hours and fan loading. Parallel fans only run at low cooling and heating loads and range from 500 to 2000 hours annually. Series fans running during all occupied and some unoccupied periods, range from 3000 to 5000 hours annually.

Parallel fans are selected to deliver 50 to 75% of design CFM and work against a lower external static pressure than series fans. Typically they operate against 0.2 to 0.3 in wg for the damper and 0.2 to 0.4 in wg for ductwork and diffusers for a total range of 0.4 to 0.7 in wg.

Dual Duct VAV Terminals

The dual duct VAV mixing boxes are used for areas needing both cooling and heating similar to VAV reheat and fan powered terminals.

Mixing boxes can deliver either 100% cold or hot supply air or mix the hot and cold air in varying proportions to meet the space loads.

Two main supply ducts serve the mixing boxes. One delivers cool air, which may be refrigerated or economizer outside air. The other duct delivers warm air, which may be centrally heated or recirculated return air. The VAV mixing box operates as follows:

Two actuators respond to the signals from the space thermostat. One controls volume output; the other controls the proportion of cold and warm air in the normal valve.

As the space cooling load decreases cold air volume drops accord-ingly. The terminal will continue to modulate the flow out of the terminal down to a minimum set value. At that point the cold air valve intake on the terminal will begin to close, and the warm air will open to a point where the higher supply air temperature will handle the new load condition. (See Figure 3-9.)

VAV Induction Units

VAV induction units are most commonly used under windows in perimeter rooms of hospitals or high-rise office buildings. Ceiling terminals in interior spaces are also used at times.

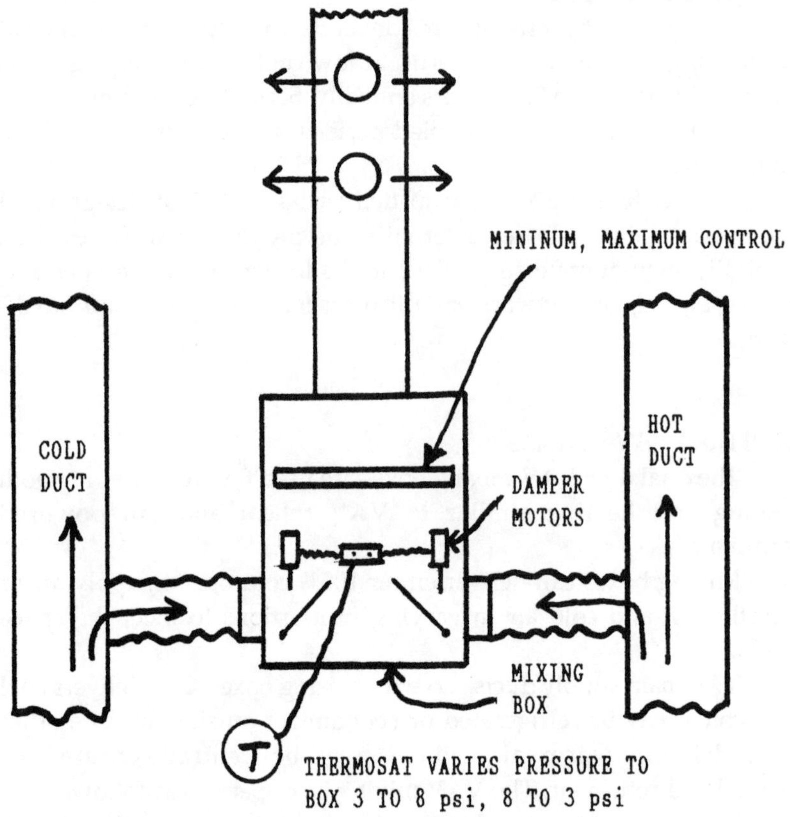

Figure 3-9. Dual Duct VAV Terminal

They mix cool primary air from the central HVAC unit with air induced from the spaces. The induction is created by primary air nozzles with small orifices that create jet streams which suck room air into the induction unit through a grille in the unit.

Cool primary air is supplied to the induction unit at a pressure high enough to overcome the induction nozzles and subsequent downstream losses.

Under full cooling, primary air dampers are wide open and induction dampers are closed. As the load falls in the space the primary air damper closes in union with the opening of the return air induction dampers maintaining substantially constant room supply air volume. At the minimum load, the induction dampers are wide open and the primary damper throttled 75 percent shut to permit the maximum induction ratio.

The induction ratio in units can vary from 25 to 65 percent. If the maximum induction of space air at full primary air CFM doesn't reduce the cooling load adequately, the primary air damper can be further closed while the induction damper remains full open.

VAV induction units may or may not contain reheat coils.

Where heating is provided, the reheat coil is activated in sequence on the further drop in the room temperature when at maximum induction and minimum primary air.

If the induction unit's primary air damper has no means of volume control, a self-contained static pressure device can be used to maintain a reasonable upstream pressure variation at a maximum of two inches.

Volume Regulator Type VAV Induction Units

Another type of VAV induction unit applied in ceilings uses volume regulators to reduce flow with a constant pressure nozzle which induces either primary air or ceiling plenum air. This function is based on a thermostat opening and closing bypass dampers in the primary air duct.

System Powered Boxes

A system powered VAV box uses air pressure from the supply air duct to power the VAV controls in the box instead of pneumatic or electrical control power.

The box itself functions as a true VAV terminal in that it throttles the air down and varies flow according to space needs.

In this process the supply duct air inflates a bellows to restrict air flow. Air is bled off from the bellows in response to a thermovalve thermostat mounted in the spaces to maintain desired flow.

A maximum volume limiter controls the selected maximum flow at the inlet which is connected to a diaphragm. Static pressure is input on the opposite side of the diaphragm which results in a velocity pressure reading. (See Figure 3-10.)

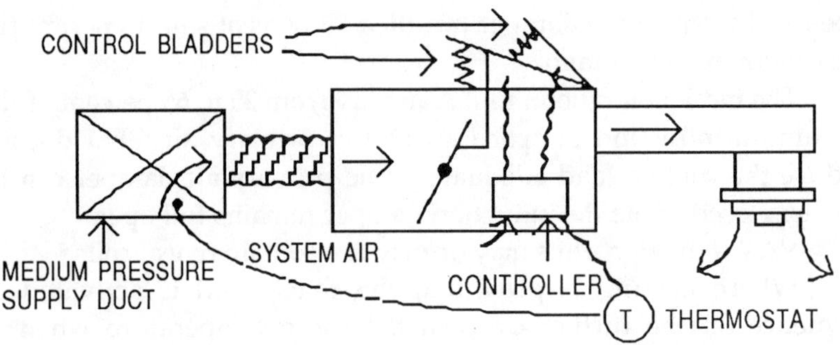

Figure 3-10. System Powered Box

Bypass Terminals

The bypass VAV box is totally different than all other types of VAV units in that it does not actually throttle or reduce the air down, rather it simply bypasses and cycles the excess air back into the return air system.

Space temperatures are maintained by varying the ratio of air entering the space and recycling it back at the terminal.

The amount of air being delivered to the box itself is constant, hence the fan volume also remains relatively constant and there are no energy savings. Also diversity cannot be taken advantage of.

Bypass boxes can be used with heating and integral coils.

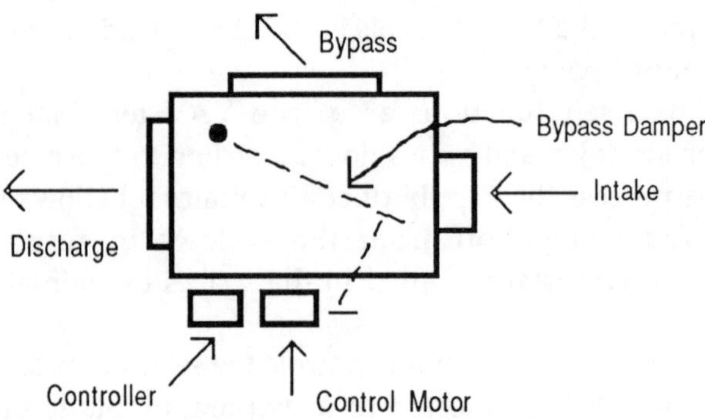

Figure 3-11. Bypass Terminal

Damper Terminals

Thermally controlled pressure dependent damper terminals are used with supply air bypasses and constant-volume fans.

As air is reduced at the terminals the excess supply air is bypassed to the main return duct or into the return air plenum ceiling.

Maximum and minimum flows through the pressure dependent damper terminals are controlled by the thermostat based on room temperature.

If reheat coils are employed they are activated when the minimum flow cannot provide heat as required.

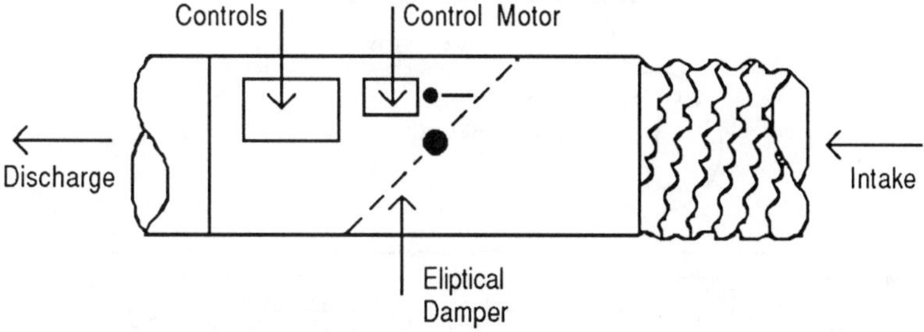

Figure 3-12. Damper Terminal

Volume Limiting

Combination pressure independent at maximum and minimum flows provides absolute flow regulation at these points, and is pressure dependent in between.

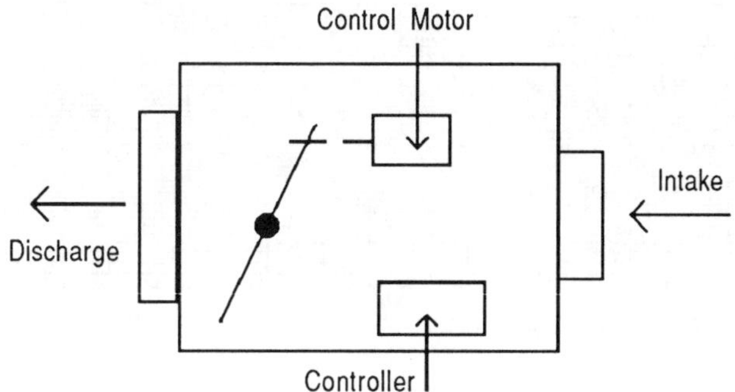

Figure 3-13. Volume Limiting Terminal

CALIBRATING PNEUMATIC VAV BOXES

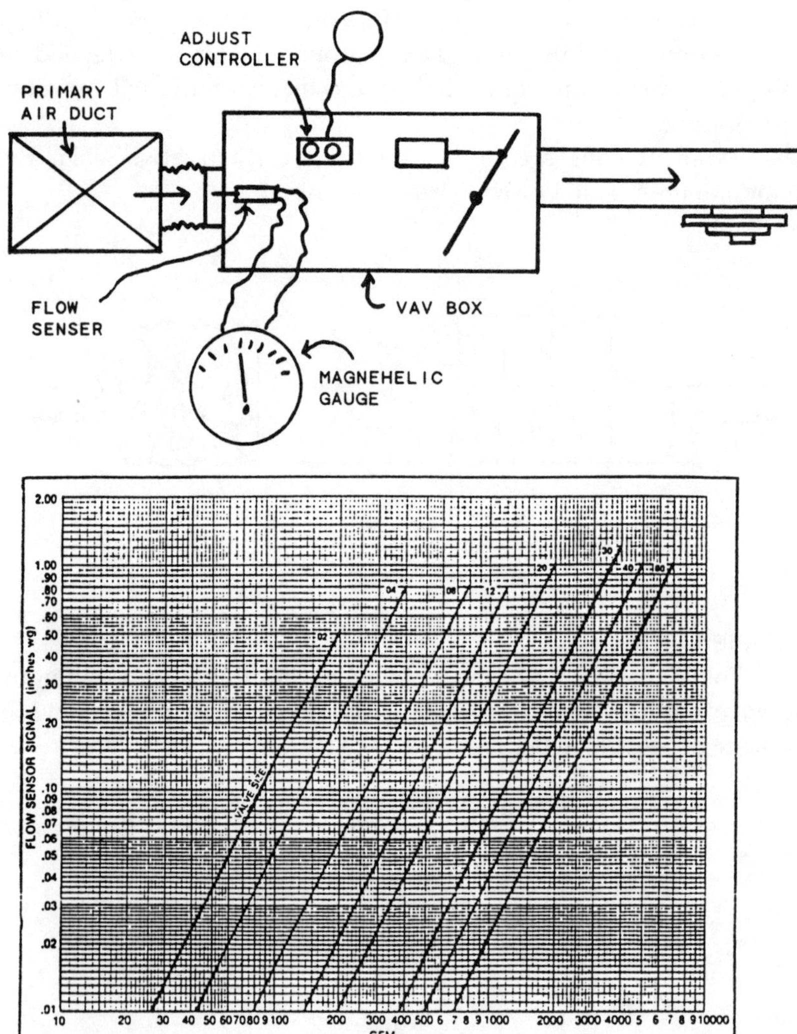

FIGURE 79 - Flow Signal/CFM Chart for Ring Type Flow Sensors

Figure 3-14. Calibrating Pneumatic VAV Boxes

CALIBRATING ELECTRIC VAV BOXES

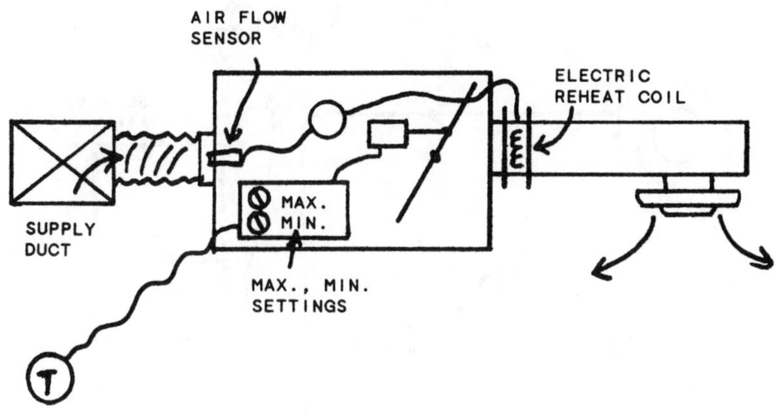

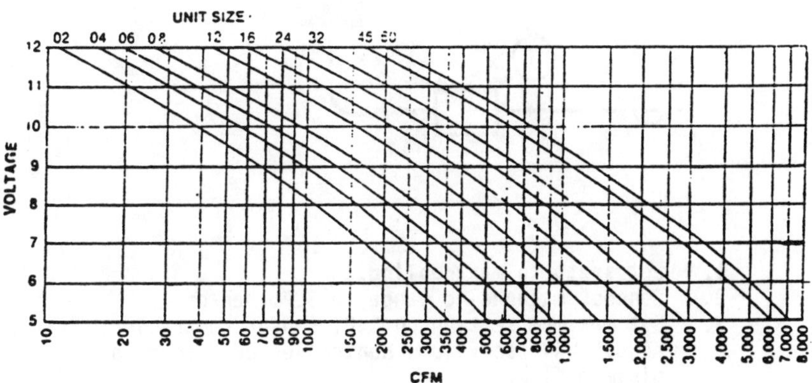

Figure 3-15. Calibrating Electric VAV Boxes

SETTING STATIC PRESSURE SENSORS IN PNEUMATIC SYSTEMS

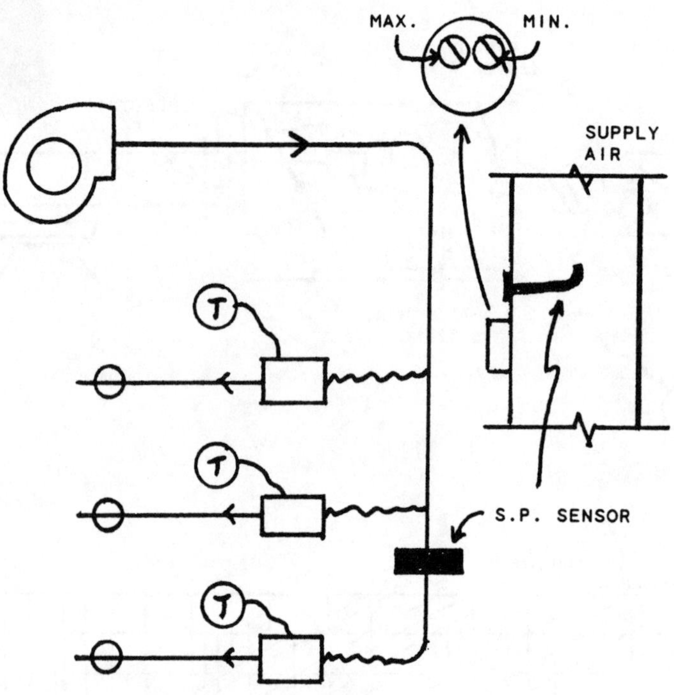

1. Set stats to max cooling.
2. Connect magnehelic gauge to S.P. sensor.
3. Set maximum setting on S.P. sensor controller.
4. Adjust VAV fan volume controller or damper to minimum flow.
5. Read S.P. at sensor for minimum.
6. Set minimum setting on S.P. sensor controller.

Figure 3-16. Setting Static Pressure Sensors in Pneumatic Systems

Chapter 4
METHODS OF VARYING FAN VOLUME

There are a number of different methods of varying fan volume in order to be proportional to the demand of the VAV terminals, to maintain proper duct pressure and to save energy.

The only true way power can be saved in a VAV system is by a reduction in BHP requirements at the fan. This is achieved by reducing the volume of air flowing through the fan.

The amount of BHP savings are contingent on a number of factors–the method of fan volume control, the effectiveness of its application, fan performance curves, and the minimum amp draw of the motor.

Fan CFM's can be reduced in a number of ways: by riding the fan curve of a forward curve fan, with intake or discharge dampers, motor inverters, automatic variable pitch sheaves, and variable pitch vaneaxial fans.

When selecting a method of fan volume control, initial costs and the amount of savings are factors in the selection and evaluation.

Fan performance must be examined as to the CFM, SP, percent minimum/maximum flows desired, the maximum and minimum load efficiencies, partial loads stability, and the effect on fan sound loads.

Suitability of the method in terms of operation, maintenance and reliability must also be considered.

METHODS OF VARYING FAN VOLUME

1. Riding the fan curve of a forward curve fan with terminal throttling

2. Riding the fan curve on a forward curve fan with a discharge damper and terminal throttling
3. Inlet vane dampers on air foil centrifugal fans
4. Motor inverters (variable frequency motor speed controllers)
5. Variable pitch vaneaxial fans
6. Variable pitch motor pulleys
7. Two-speed eddy current coupling
8. Automatic centrifugal fan wheel shroud
9. Fan bypass

PERCENT HORSEPOWER SAVED

The percent break horsepower saved varies with the type of fan volume control and the minimum BHP draw possible on the particular fan motor.

The motor inverter, which reduces motor and fan rpms, saves the highest percentage and is the most efficient. Second in line is a forward curve fan with inlet vanes. The method with the least savings, but nevertheless still of great value, is the backward incline centrifugal fan with a discharge damper. (See Fig. 4-1)

MINIMUM MOTOR AMP DRAWS

There is a certain minimum amount of energy required to run a constant rpm motor at its rated speed regardless if there is an external load on the motor or not.

The higher the horsepower, the lower the minimum energy draw required. Motors from 20 HP to 100 HP generally can go down to 40% to 35% of full load amps. Motors in the range of 5 HP to 20 HP can generally reduce to 60% to 40% of full load amps. One to 5 HP motors can reduce to around 60% to 80% and fractional HP motors save negligible amounts of energy.

If a centrifugal fan motor is only running at 50% of its load already, there would be very little if any energy savings made available on the motor by adding a fan volume control device, if the device is a constant rpm type such as a damper or riding the fan

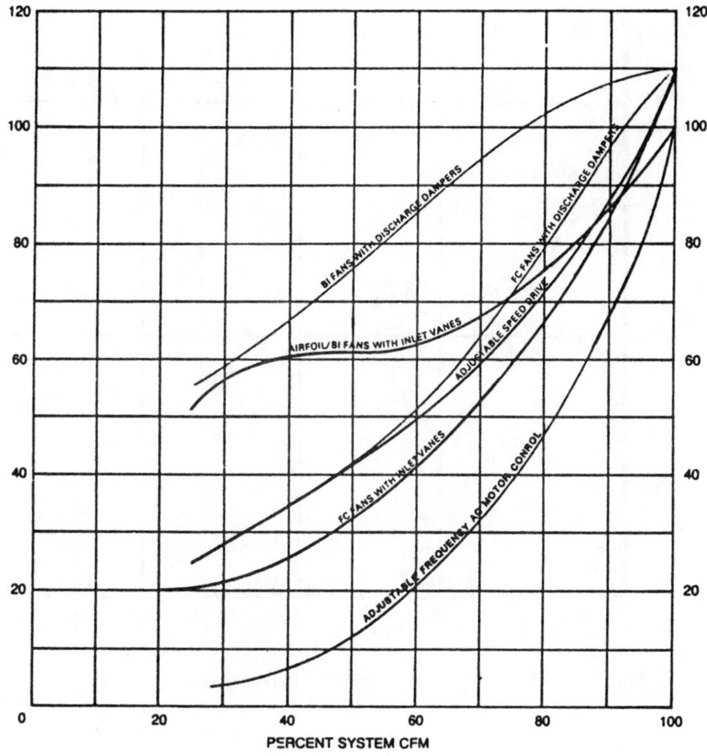

Figure 4-1. The above graph shows the maximum BHP savings on the fan motor in relationship to the percent of system CFM for various methods of reducing fan volume.

curve. An inverter or some other method of reducing motor rpms would be required for further gain.

The closer the actual load on the motor is to the full load amps, the greater the potential savings with a fan volume controller. (See Figures 4-2 and 4-3.)

RIDING FAN CURVE WITH A FORWARD CURVE FAN AND TERMINAL THROTTLING

For smaller type VAV systems, either in the 1/2- to 2-inch system SP range or in the 2- to 4-inch range, where the forward

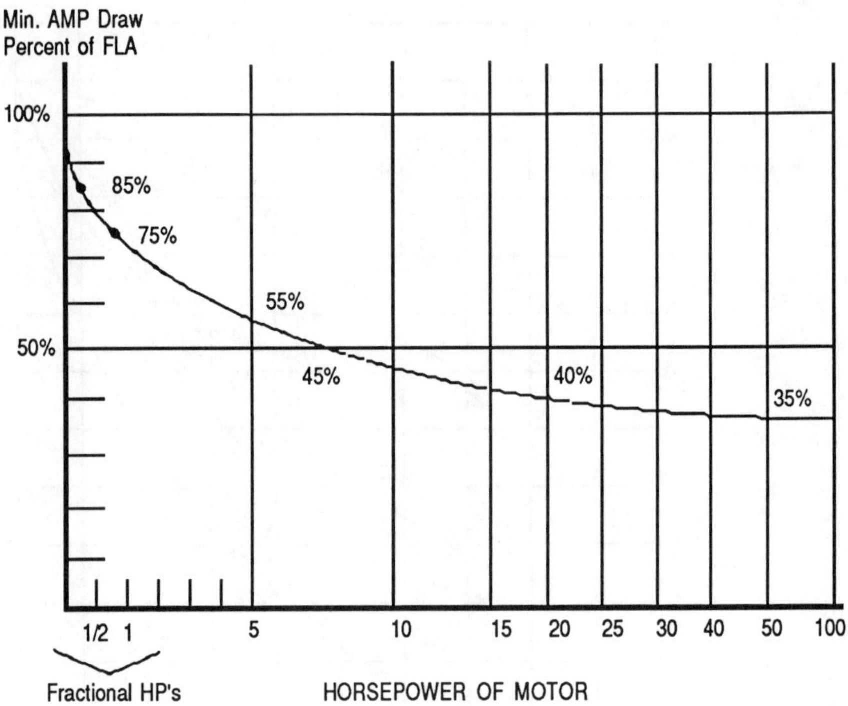

Figure 4-2. Minimum motor amp draws as a percent of full load amps.

curve fan is most efficient, riding the fan curve as terminals modulate is an easy and most economical choice.

Not only are forward curve fans the lowest in first cost, but they also are relatively stable over a wide range of CFM's and they tolerate air flow changes with a relatively small effect on system static pressure.

Air volume and BHP on forward curve fans reduced automatically as a result of system resistance increases when terminals throttle. (See Figure 4-4.)

FACTORS AFFECTING FAN OPERATING COSTS

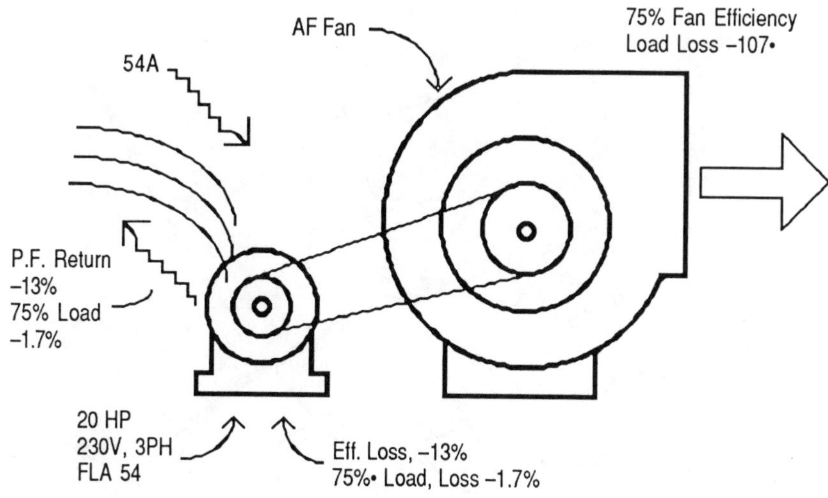

	Precentage Load		
	100%	75%	50%
Motor Efficiency	87%	83%	79%
Power Factor	.87	.83*	.80
Fan Efficiency	85%	80%	60%

Operates 1/2 time @ 75% load
8¢/kWh, 15HP x $400/HP/YR = $6000/YR

Figure 4-3. As the percentage load on a motor and fan drops, both the motor and fan efficiencies simultaneously reduce.

RIDING THE FAN CURVE WITH FORWARD CURVE FANS AND DISCHARGE DAMPERS

Riding the constant rpm fan curve on forward curve fans with discharge dampers is very applicable for higher ranges of static pressure, and where sound levels are important.

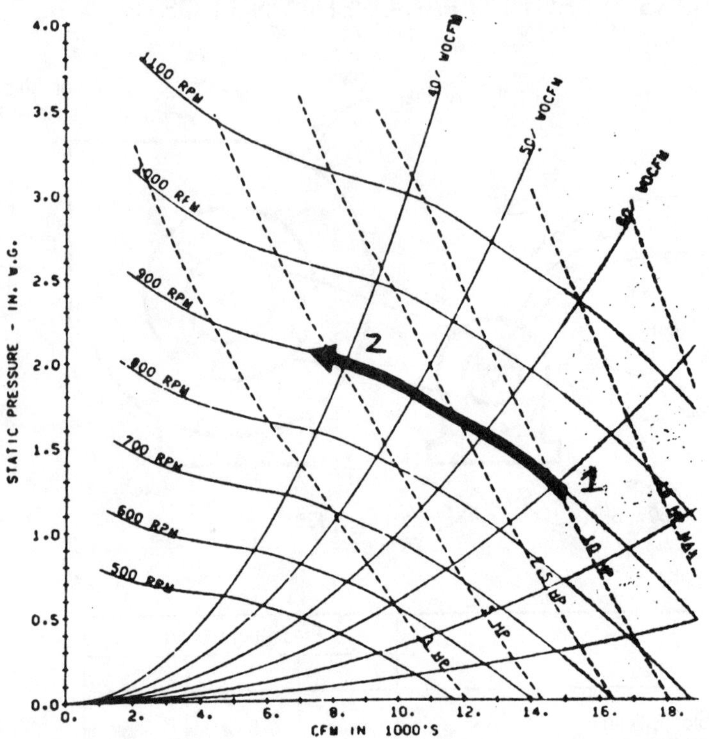

Figure 4-4. Constant rpm curve illustrates how CFM reduces from 16,000 to 9,000 and BHP reduces from 10 to 5 as terminals throttle down and increases the system resistance from 1.2 to 2.0 inches static pressure.

When load demand is reduced in the spaces, terminals throttle down, duct pressure increases, the duct pressure sensor detects and sends message to discharge damper to throttle proportionally according to setting. (See Figure 4-5.)

INLET VANE DAMPERS ON AIR FOIL FANS

Inlet vane dampers on air foil fans are used in systems with high CFM's and static pressures. Unlike the forward curve fan which allows overloading and riding the fan curve, the air foil fan is non-overloading and doesn't ride the fan curve or respond to discharge dampers properly.

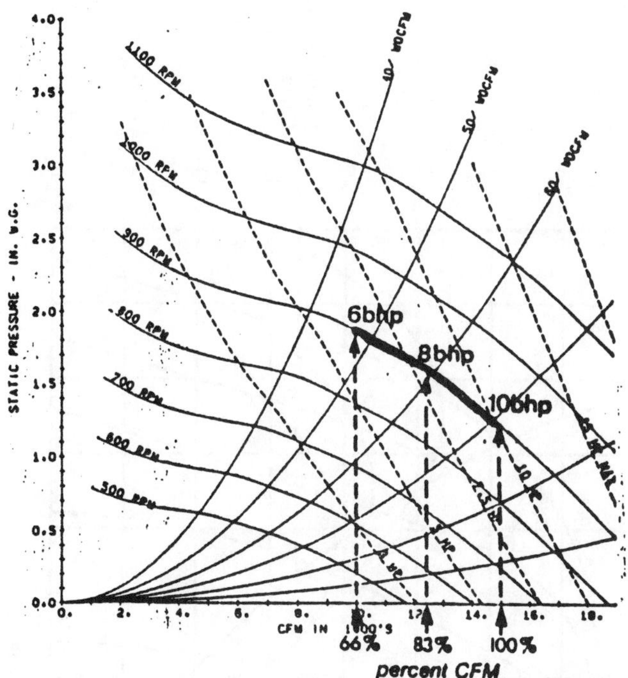

Figure 4-5. The above curve shows that a forward curve fan with a discharge damper rides the constant rpm fan curve. It draws 10 BHP at 100 percent flow at a constant rpm, 8 BHP if the damper is throttled to an 83 percent flow and, 6 BHP at 66 percent flow.

Hence, inlet vanes are installed on air foil fans, adjusting the air flowing into the fan according to system air volume and pressure variations.

Air spins in the direction of fan rotation at the inlet of the fan which results in less static pressure and horsepower being required. The vanes not only reduce the air flow but also change the fan system curve. There is a different fan performance curve at each damper setting. (See Figure 4-6.)

MOTOR INVERTERS

Variable frequency AC motor speed controllers ("inverters") reduce the motor rpm and consequently the fan rpm. This results in the

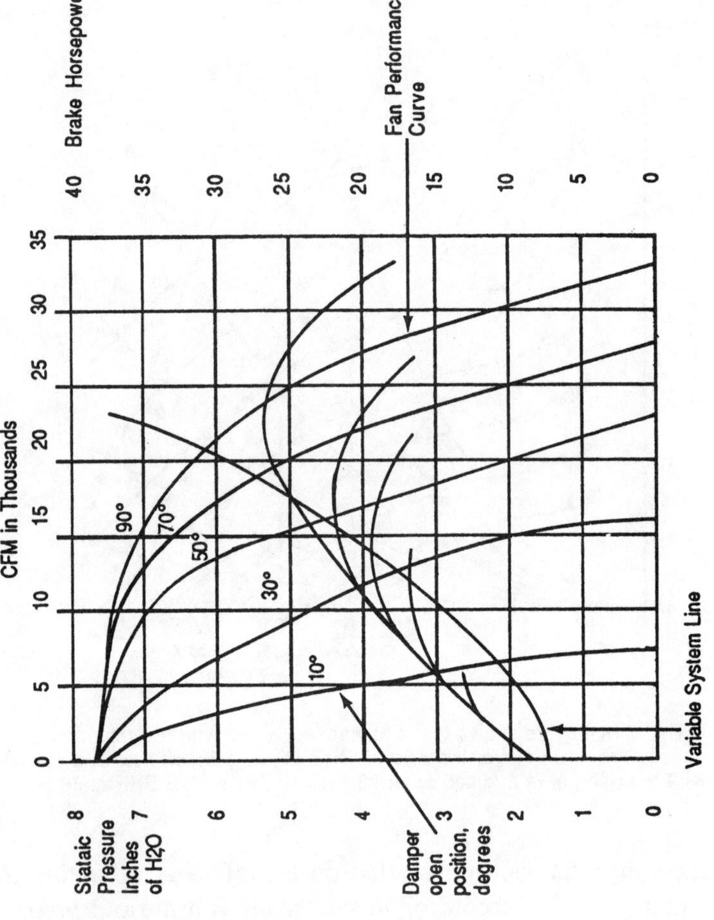

Figure 4-6. Curves illustrate how different degrees of damper openings change system line and fan performance curve. A 90-degree open damper allows 23,000 CFM at about 5-1/2 inch S.P. and draws 28 brake horsepower. Closing the damper to a 50-degree position throttles air flow down to 14,000 CFM at 3.9 inches SP and draws only about 20 brake horsepower. This means a 40% CFM reduction reduces BHP 29 percent.

most efficient method of fan air volume control and highest energy savings of all volume control methods.

Incoming power is first inverted in the unit to DC by thermistors, then the frequency is changed and finally thermistors invert them back to AC current again to drive the AC motor at a lower rpm.

FREQUENCY INVERTERS NOTES

1. Inverter maintains motor efficiency regardless of rpm or horsepower.

2. Motors don't lose their power factors and stay at about 95%.

3. A 10 percent increase in temperature on a motor decreases insulation life 50%.

4. If there are noise interference problems from computers or programmable controllers, filter noise out with a isolation transformer.

5. One inverter can be used for 2 or more motors if the motors are running at the same speed and starting and stopping times are the same. Add amps of motors up to select inverter size.

6. Transducers put out 0 to 10 volts or 4 to 40 milliamps.

(See Figure 4-7.)

VARIABLE PITCH VANEAXIAL FANS

The volume of air changes with the angle of the blades in a constant rpm variable pitch vaneaxial fan. This results in not only less air, but also a reduction in static pressure and brake horsepower. Vaneaxial fans with variable pitched blades are good for larger applications generally up to 60,000 CFM and where higher percentage throttling is needed. There are no drive losses with variable pitch vaneaxial fans, which reduces operational costs. However initial costs are higher. (See Figure 4-8.)

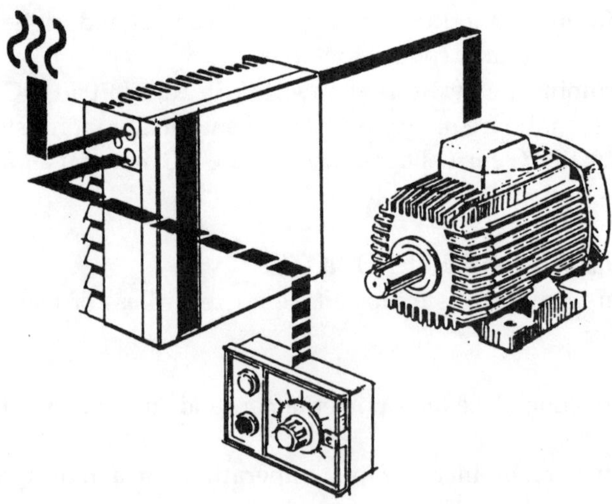

Figure 4-7. Frequency inverters reduce the motor rpm's and consequently the fan speed and flow.

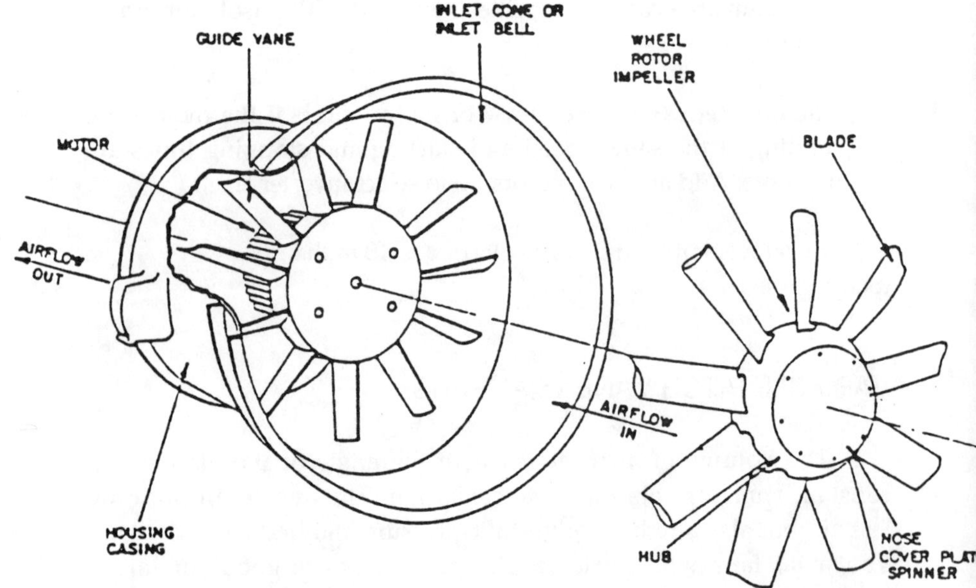

Figure 4-8. The two-part graph above shows the CFM and static pressure at different blade angles in the top graph, and below the BHP versus the blade angle is depreciated. After the operating point is established on the top graph a line is projected vertically straight down and intersects the same blade angle curves below. Then a horizontal line is projected to the left for the resultant BHP.

VARIABLE PITCH MOTOR SHEAVES

Automatic variable pitch motor sheaves with special gear boxes and belts change the fan rpm within a certain range. This results in reduced CFM, static pressure and BHP according to the fan load. They do not have the full range of the motor frequency inverter.

Figure 4-9.

TWO-SPEED MOTORS AND EDDY CURRENT DRIVES

Two-speed motors or eddy current drives are best suited for use with forward curve fans only. They should not generally be used with air foil fans and are not energy effective with them.

When two-speed operation is applied to a forward curve fan, the switching point from the high to the low speed should be carefully chosen by analyzing the two rpm curves in regard to static pressure and BHP changes.

The two-speed application is best controlled with measurement of air flow rather than static pressure.

Eddy current drives use standard ac motors and eddy current couplings. The input rotor (constant speed) generates a rotating magnetic flux for the output rotor.

This flux generates eddy currents in the output rotor when differences in output and input rotor speeds exist. The eddy currents generate magnetic fields that are attracted to the originating field, resulting in an output torque.

The electronic circuitry of eddy current drives is considerably less complex than dc or ac drives, and their proponents cite the familiarity and extensive use history of eddy current drives as advantages. Eddy current drives are less expensive than the ac and dc drives, but also are less efficient.

INTERNAL WHEEL SHROUDS

Internal Centrifugal Fan Shrouds

The volume of a centrifugal fan can be controlled with cylindrical dampers which automatically slide over the air foil or backward inclined wheel reducing the effective widths of the wheel which decreases cfm and static pressure.

This method reduces power consumption because of different fan curves and lower statics. The fan runs at a constant rpm and motor efficiencies are not realized. (See Figure 4-10.)

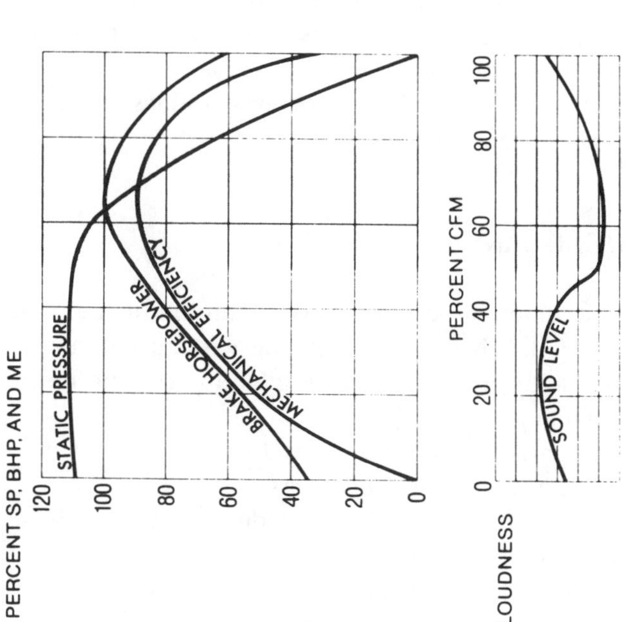

Figure 4-10. Internal Wheel Shroud.

FAN BYPASS

In a fan bypass air is cycled in the fan housing with an external bypass fitting. Fan bypasses **do not** provide any net savings in power since the fan is still running at the same rpm and conveying the same total volume of air through it. They only function as variable volume controllers.

GENERAL INFORMATION ON FANS

The fan is the heart of the air distribution system. It pumps the blood to all the parts of the body. The function and understanding of the fan and air distribution system is as important to the balancer as the heart and circulatory system are to a human.

There are numerous types of fans:

- **Centrifugals** with four different types of wheels:
  - Backward inclined
  - Forward Curve
  - Air Foil
  - Radial Paddle Wheel
- **Tubular incline fans** with centrifugal wheels
- **Tubular axial fans** with or without straightening vanes
- **Propeller fans**
- **Roof exhaust fans** of various configurations with centrifugal or propeller wheels

CENTRIFUGAL FANS

The most widely used fan for HVAC and industrial systems is the centrifugal fan. It has a scroll housing with air entering a round inlet on one side or both sides, turning 90 degrees and discharging through a rectangular opening.

The main parts of a centrifugal fan (See Figure 4-11) are:

- Scroll housing
- Wheel with hub
- Inlet cone
- Cut off plate
- Shaft
- Bearings
- Motor pedestal or bearing base
- Motor
- Drives
- Supports

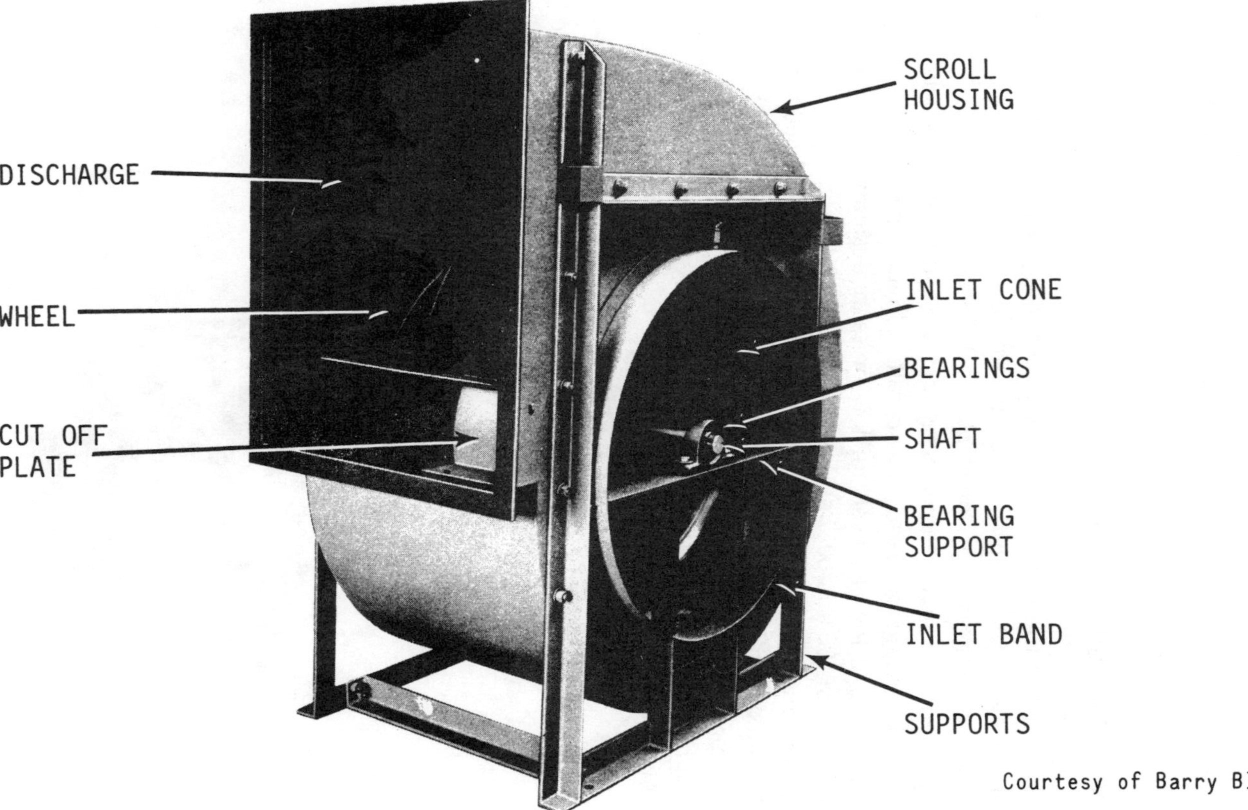

Figure 4-11. Centrifugal fan, single width single inlet (SWSI) with airfoil wheel motor arrangement, where the motor gets mounted on floor detached from fan.

The acceptable rpm range of centrifugal fans varies with the type of wheel, diameter of the wheel and the recommended top speed, which are in turn predicted on appropriate outlet velocities.

Types of Centrifugal Wheels

The four types of centrifugal wheels are:

1. **Backward Inclined (BI):** The flat blades around a BI wheel are inclined away from the direction of air flow. The wheels are non-overloading and provide stable air delivery.
   They operate at higher efficiencies than forward curve wheels but are not as quiet because they operate at higher speeds.
   The rpm range for BI wheels for low pressure systems may run roughly from 400 to 1700 rpm in actual practice.
   The rpm range for typical high pressure systems is roughly 600 to 1500 rpm depending on the diameter of the wheel, etc.
   Static pressure range is from roughly 1/2 to 2 inches for low pressure systems; 2 to 8 inches for medium pressure systems; and over 6 inches for higher pressures systems.

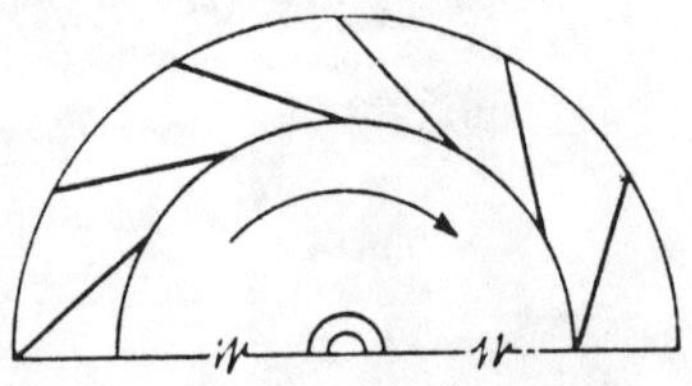

Figure 4-12.

2. **Air Foil Wheels (AF):** Air foil wheels are similar to BI wheels in that they slope away from the direction of air flow, but their shape (instead of being flat as the BI) is similar to the wings of an airplane.
   They have the similar performance characteristics as BI wheels also, except they are more efficient and run at slightly higher top speeds to deliver the same volume of air. Consequently rpm ranges are slightly higher also. They are used more extensively for higher volume and high pressure systems.

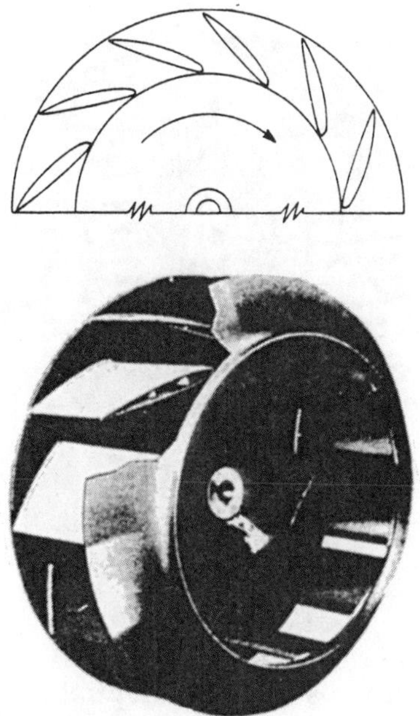

Figure 4-13.

3. **Forward Curve Wheels (FC):** The blades of a forward curve wheel are much narrower than BI or FC wheels and are curved like a crescent. The inside of the curve faces and is slightly inclined in the direction of air flow. The wheel accelerates the air and discharges it at a higher speed than the fan is rotating. Of the three, BI, AF and FC, the FC is the least efficient.

FC wheels move large masses of air at low rpm's quietly and require less space.

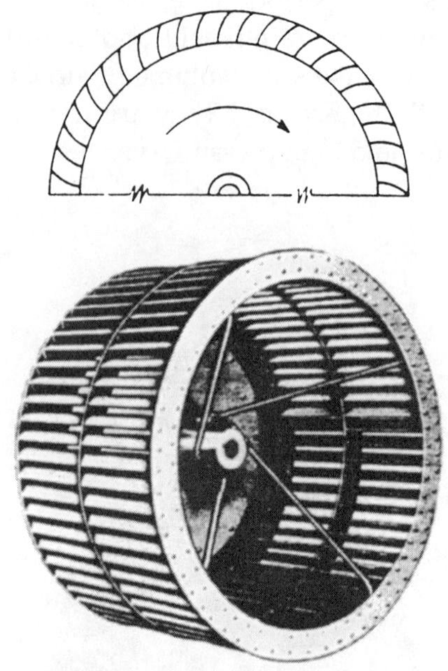

Figure 4-14.

FC's are normally used in residential and light commercial systems and for light duty exhaust where maximum air delivery and low noise levels are required. Not recommended where dust or fumes would adhere to blades.

4. **Radial Blade Wheel:** Has straight blades which are largely self cleaning and of great structural strength making them suitable for industrial exhaust applications, abrasive dusts, materials handling, fumes containing particles, grease, acids, etc. They can stand high speeds and pressures but are noisier.

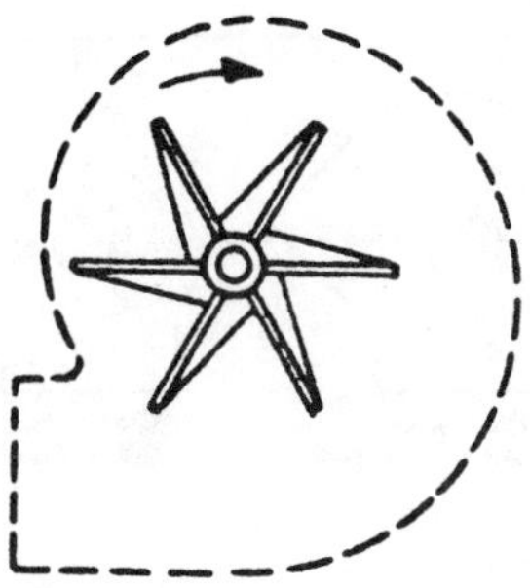

Figure 14-15.

Widths and Inlets

Centrifugal fans are constructed either as single width single inlet (SWSI) or double width double inlet (DWDI). The SWSI fan has an inlet only on one side, opposite the drive side, whereas the DWDI has two inlets, one on each side and the discharge is about 75 percent wider than the SWSI.

Rotations and Discharges

The rotation and discharge of a centrifugal fan are always viewed from the drive side, Fig. 4-16. Centrifugal fans can be discharged in 8 basic directions, relative to the rotation, horizontally with the discharge at the top or bottom, vertically up or down, or at a 45-degree angle upwards or downwards.

That means there can be two top horizontal discharges, one with the wheel turning clockwise, and the other counter-clockwise.

Discharges are abbreviated using the first letters: top horizontal TH, bottom horizontal BH, up blast UB, top angular up TAU, etc. Rotations are abbreviated CW and CCW, clockwise and counter clockwise.

Motor and Drive Arrangements

There are various arrangements for motors and drives on centrifugal fans and AMCA has assigned them number designations, Fig. 4-18. Motors can be mounted inside the base, arrangement 10, and can be belt driven. This arrangement is generally called a utility or vent set.

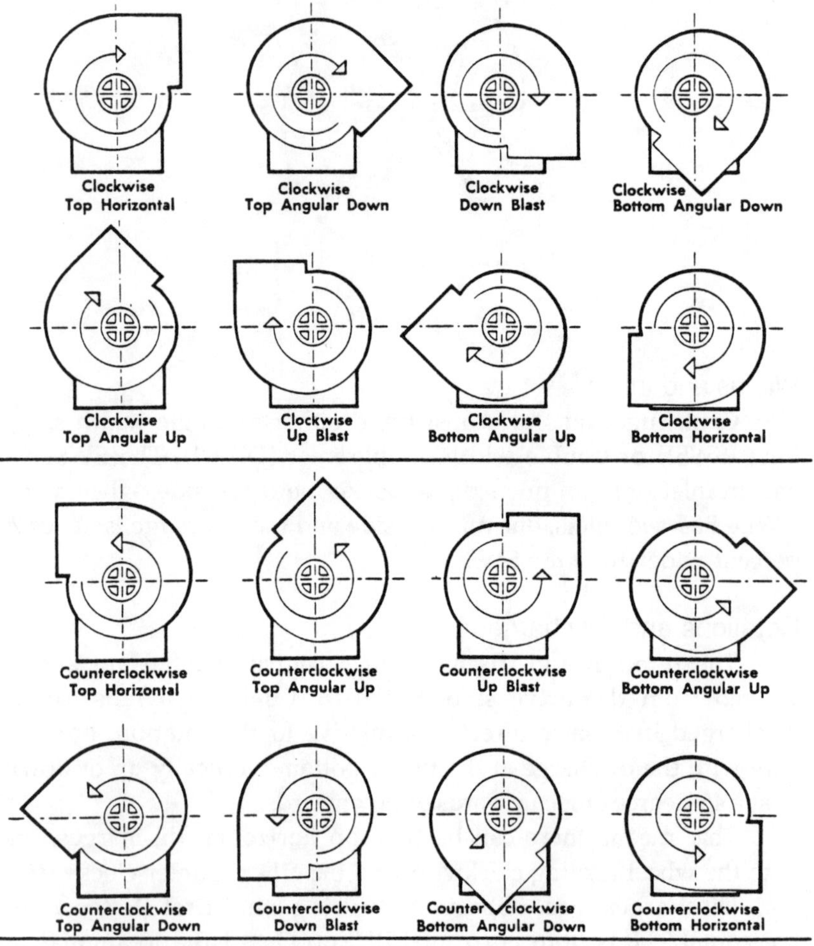

Figure 4-16. AMCA rotation and discharge designations.

A very common arrangement in HVAC systems is arrangement 3. The bearings are mounted on both sides of the fan, the shaft overhung on the drive side for a belt drive and the motor mounted on an integral isolation base on the floor.

When a motor is mounted on the floor with a centrifugal fan, as with arrangement 3, it can be positioned to the right or left, outside or under the housing. These positions are designated W, X, Y, Z (Figure 4-17).

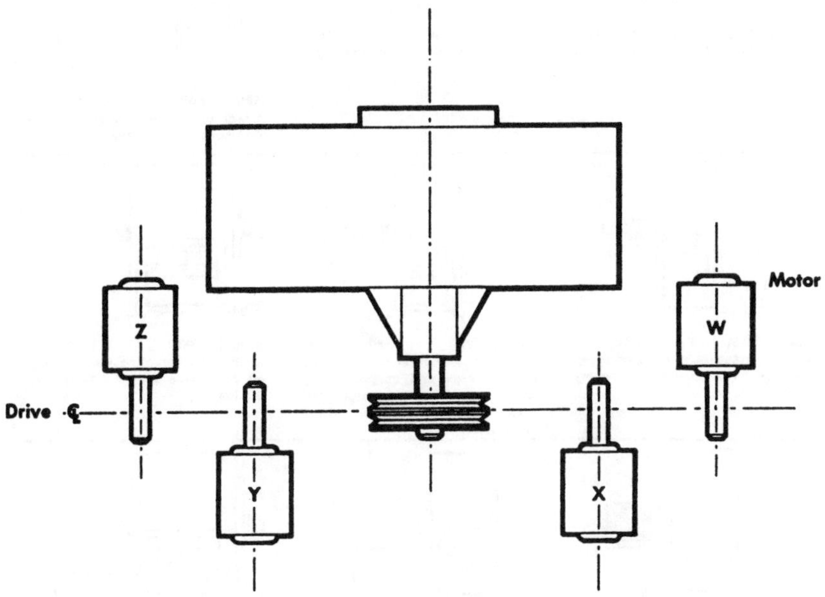

Figure 4-17. Plan View of Centrifugal Fan.

Classes of Construction

AMCA has developed four classes of centrifugal fan construction in standards No. 2408-60, for gauges of metal etc., based on the static pressure rating of the fan and the outlet velocity at standard air conditions. For example, a Class I SWSI fan can go up to a maximum of 5 inches static pressure at a 2300 fpm outlet velocity, or 2-1/2 inches at 3200 fpm outlet velocity. Class II has higher ratings and Class III higher yet.

There are maximum rpms for SWSI, and DWDI fans for the different classes of fans. For example a 20-inch-diameter SWSI fan in Class I has a maximum rpm of about 2080, in Class II 2700 rpm, and in Class III 3410 rpm. A 44-inch-diameter fan has maximums of 882, 1150 and 1447 rpms respectively.

DRIVE ARRANGEMENTS FOR CENTRIFUGAL FANS

SW - Single Width DW - Double Width

SI - Single Inlet DI - Double Inlet

Arrangements 1, 3, 7 and 8 are also available with bearings mounted on pedestals or base set independent of the fan housing.

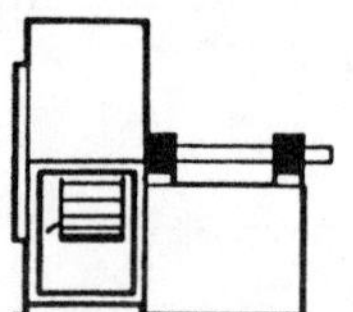

ARR. 1 SWSI For belt drive or direct connection. Wheel overhung. Two bearings on base.

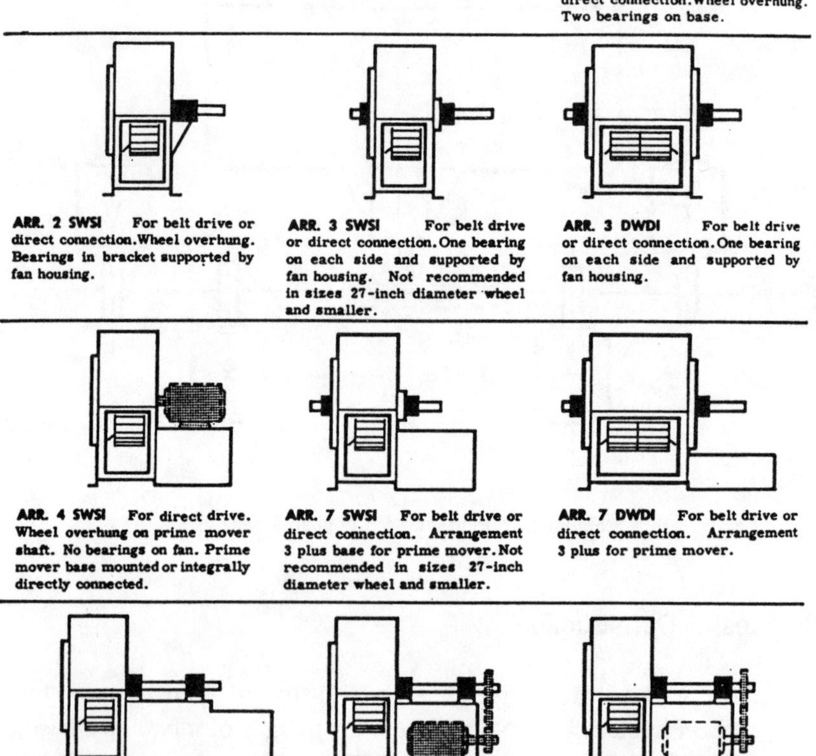

ARR. 2 SWSI For belt drive or direct connection. Wheel overhung. Bearings in bracket supported by fan housing.

ARR. 3 SWSI For belt drive or direct connection. One bearing on each side and supported by fan housing. Not recommended in sizes 27-inch diameter wheel and smaller.

ARR. 3 DWDI For belt drive or direct connection. One bearing on each side and supported by fan housing.

ARR. 4 SWSI For direct drive. Wheel overhung on prime mover shaft. No bearings on fan. Prime mover base mounted or integrally directly connected.

ARR. 7 SWSI For belt drive or direct connection. Arrangement 3 plus base for prime mover. Not recommended in sizes 27-inch diameter wheel and smaller.

ARR. 7 DWDI For belt drive or direct connection. Arrangement 3 plus for prime mover.

ARR. 8 SWSI For belt drive or direct connection. Arrangement 1 plus extended base for prime mover.

ARR. 9 SWSI For belt drive. Wheel overhung, two bearings, with prime mover outside base.

ARR. 10 SWSI For belt drive, wheel overhung, two bearings, with prime mover inside base.

Courtesy of Air Moving and Control Association, Inc.

Figure 4-18. AMCA motor and drive arrangements for centrifugal fans.

VANEAXIAL FANS

Axial fan wheels are housed in tubes just as the centrifugal inline fan wheel in, but the wheel is a turbine-propeller type and the air flows straight through, rather than turning 90 degrees as it does in a centrifugal wheel.

There are more blades on a vaneaxial wheel than on a propeller wheel and they are either flat or curved, of single thickness or airfoil. There may or may not be air guide vanes in the housing after the wheel on the discharge end.

Vaneaxial fans may have adjustable air foil blades which can increase or decrease the CFM by changing the pitch of the blades. The vaneaxial fan works well in VAV systems. (See Figure 4-19.)

FAN PERFORMANCE

Fan manufacturers publish performance charts for the various sizes and types of fans they produce based on tests and curves. The charts are used for selection of fans and trouble shooting problems in balancing, performance, etc.

The charts are based on the following information:

- CFM
- Outlet Velocity
- Static Pressure
- RPM
- Brake Horsepower

See Figure 4-20.

A typical chart for a particular size centrifugal fan might cover performance from 1/4 inch to 13 inches static pressure, from 400 rpm to 3000 rpm and cover the various classes.

Figures 4-21 and 4-22 are examples of performance charts for Barry airfoil fans. The first chart is for a SWSI fan and the second a DWDI fan. Both have the same size 24-1/2-inch-diameter wheels.

In Chapter 10 on air test reports you will find that supply fan S-2 is a DWDI No. 7245 Barry air foil fan. The line marked on the chart in Figure 4-21 is for 9936 CFM at 1600 fpm outlet velocity against 1-1/2 inches S.P., requires 999 rpm and draws 3.53 brake horsepower.

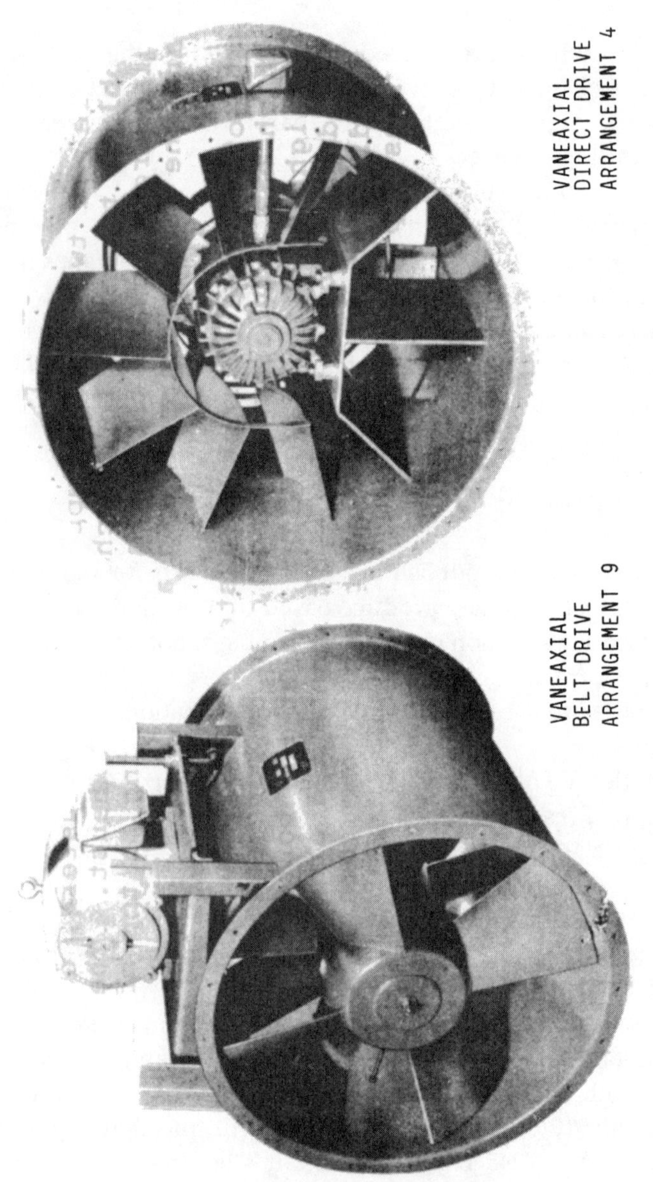
VANEAXIAL
DIRECT DRIVE
ARRANGEMENT 4
VANEAXIAL
BELT DRIVE
ARRANGEMENT 9

Figure 4-19.

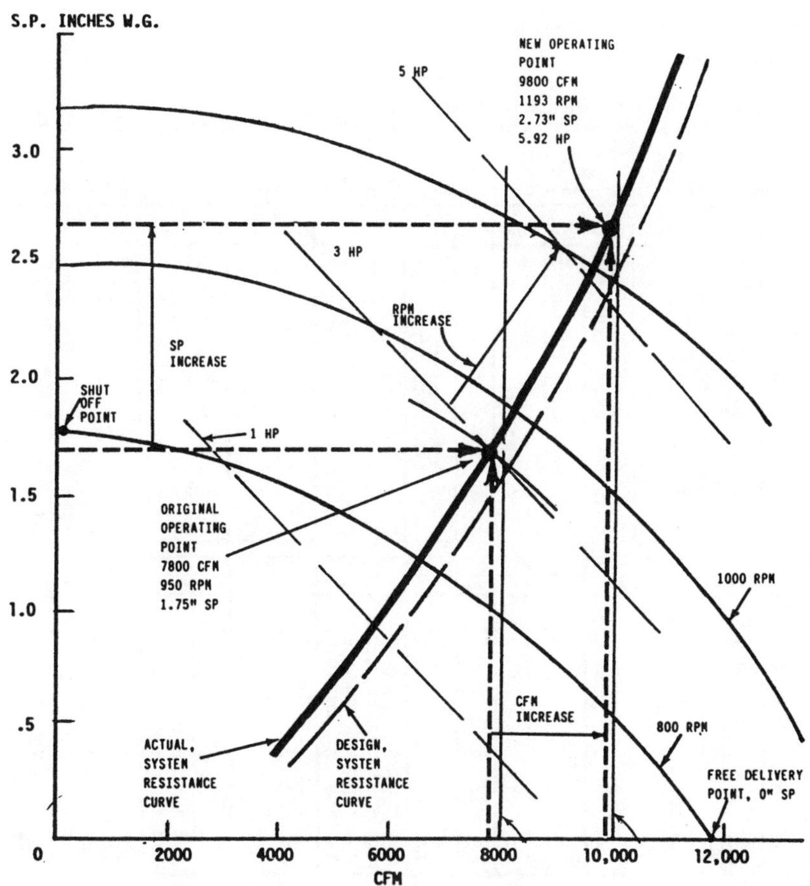

Figure 4-20. Fan performance curve showing how operating point moves on system resistance line when rpm is increased.

Fan Curves

There are two types of fan curve graphs. The basic fan curve plots the CFM against the resistance of the system in static pressure, and plots rpm, system resistance and BHP lines.

The second graph plots percentage CFM from shutoff to free delivery against percentage static pressure and has efficiency lines plotted on it. This is called a performance characteristic curve.

CENTRIFUGAL AIR FOIL FAN, No. 7245

DOUBLE WIDTH, DOUBLE INLET

INTAKE AREA = 7.280 SQ. FT. • WHEEL DIAMETER 24-1/2"
OUTLET AREA = 6.21 SQ. FT. • TIP SPEED F.P.M. = 6.41 x R.P.M. • MAXIMUM B.H.P. = $3.550\left(\frac{R.P.M.}{1000}\right)^3$

Top white area, Class I / Top grey area, Class II / Bottom white area, Class III / Bottom grey area, Class IV

C.F.M.	O.V.	1/4" S.P.		3/8" S.P.		1/2" S.P.		5/8" S.P.		3/4" S.P.		7/8" S.P.		1" S.P.		1-1/4" S.P.		1-1/2" S.P.	
		R.P.M.	B.H.P.	R.P.M.	B.H.P.	R.P.M.	B.H.P.	R.P.M.	B.H.P.	R.P.M.	B.H.P.	R.P.M.	B.H.P.	R.P.M.	B.H.P.	R.P.M.	B.H.P.	R.P.M.	B.H.P.
4968	800	451	0.33	499	0.44	543	0.56	585	0.69	624	0.82	662	0.95	698	1.09	767	1.39		
5589	900	483	0.40	528	0.52	570	0.65	609	0.79	646	0.93	681	1.07	716	1.22	781	1.54	843	1.87
6210	1000	517	0.48	559	0.62	598	0.76	635	0.91	671	1.06	704	1.21	737	1.37	799	1.70	858	2.04
6831	1100	551	0.58	591	0.73	629	0.88	664	1.04	697	1.20	729	1.36	761	1.53	820	1.88	876	2.24
7452	1200	587	0.69	625	0.85	660	1.02	694	1.19	726	1.36	756	1.53	786	1.71	843	2.08	897	2.46
8073	1300	623	0.81	659	0.99	693	1.17	725	1.35	755	1.53	785	1.72	813	1.91	868	2.29	920	2.69
8694	1400	659	0.96	694	1.14	726	1.33	757	1.53	786	1.72	815	1.92	842	2.12	895	2.53	945	2.95
9315	1500	697	1.12	730	1.32	761	1.52	790	1.73	818	1.93	846	2.14	872	2.36	923	2.79	971	3.23
9936	1600	734	1.30	766	1.51	795	1.72	824	1.94	851	2.16	877	2.38	903	2.61	952	3.06	999	3.53
10557	1700	772	1.50	802	1.72	831	1.95	858	2.18	885	2.41	910	2.65	935	2.88	982	3.36	1027	3.85
11178	1800	810	1.72	839	1.96	867	2.20	893	2.44	919	2.69	943	2.93	967	3.18	1013	3.68	1057	4.19
11799	1900	849	1.96	877	2.22	903	2.47	929	2.72	953	2.98	977	3.24	1000	3.50	1045	4.03	1087	4.56
12420	2000	888	2.23	915	2.50	940	2.76	965	3.03	989	3.30	1012	3.57	1034	3.85	1077	4.40	1119	4.95
13662	2200	966	2.85	991	3.14	1015	3.43	1038	3.73	1060	4.02	1082	4.32	1103	4.61	1144	5.21	1183	5.82
14904	2400	1045	3.58	1068	3.90	1091	4.22	1112	4.53	1133	4.85	1154	5.17	1174	5.50	1213	6.15	1250	6.80
16146	2600	1124	4.44	1146	4.78	1167	5.12	1188	5.46	1208	5.81	1227	6.16	1246	6.50	1283	7.20	1318	7.91
17388	2800	1204	5.43	1225	5.79	1245	6.16	1264	6.53	1283	6.90	1301	7.27	1319	7.64	1354	8.39	1388	9.15
18630	3000	1285	6.56	1304	6.95	1323	7.34	1341	7.74	1359	8.13	1376	8.53	1394	8.93	1427	9.73	1460	10.54
19872	3200	1365	7.85	1384	8.26	1401	8.68	1419	9.10	1436	9.52	1452	9.94	1469	10.37	1501	11.22	1532	12.08
21114	3400	1446	9.30	1463	9.74	1480	10.18	1497	10.63	1513	11.08	1529	11.52	1545	11.97	1575	12.87	1605	13.78

C.F.M.	O.V.	2" S.P.		2-1/2" S.P.		3" S.P.		3-1/2" S.P.		4" S.P.		4-1/2" S.P.		5" S.P.		5-1/2" S.P.		6" S.P.	
		R.P.M.	B.H.P.	R.P.M.	B.H.P.	R.P.M.	B.H.P.	R.P.M.	B.H.P.	R.P.M.	B.H.P.	R.P.M.	B.H.P.	R.P.M.	B.H.P.	R.P.M.	B.H.P.	R.P.M.	B.H.P.
7452	1200	999	3.26	1094	4.13	1184	5.05												
8073	1300	1018	3.53	1110	4.43	1196	5.38	1279	6.38										
8694	1400	1039	3.83	1128	4.75	1211	5.73	1292	6.76	1369	7.84								
9315	1500	1062	4.15	1148	5.11	1229	6.11	1307	7.17	1381	8.27	1453	9.42						
9936	1600	1087	4.49	1170	5.48	1248	6.53	1324	7.61	1396	8.74	1466	9.92	1534	11.14				
10557	1700	1113	4.85	1193	5.89	1270	6.97	1343	8.09	1413	9.25	1482	10.45	1548	11.69	1612	12.98	1675	14.31
11178	1800	1140	5.24	1218	6.32	1292	7.43	1363	8.59	1432	9.78	1499	11.01	1563	12.28	1626	13.60	1687	14.95
11799	1900	1168	5.65	1244	6.77	1316	7.93	1386	9.12	1453	10.35	1517	11.61	1580	12.92	1641	14.26	1701	15.63
12420	2000	1197	6.09	1271	7.26	1342	8.46	1409	9.69	1474	10.95	1537	12.25	1599	13.58	1659	14.95	1717	16.36
13662	2200	1258	7.05	1328	8.31	1395	9.60	1459	10.91	1522	12.25	1582	13.63	1640	15.03	1697	16.47	1753	17.93
14904	2400	1321	8.13	1388	9.48	1452	10.86	1513	12.26	1573	13.69	1631	15.14	1687	16.62	1741	18.13	1795	19.67
16146	2600	1386	9.34	1450	10.78	1511	12.25	1570	13.74	1627	15.26	1683	16.80	1737	18.36	1789	19.95	1841	21.56
17388	2800	1453	10.68	1514	12.22	1573	13.78	1630	15.37	1685	16.97	1738	18.60	1790	20.25	1841	21.92	1890	23.61
18630	3000	1522	12.16	1581	13.80	1637	15.46	1692	17.14	1745	18.84	1796	20.56	1846	22.30	1895	24.05	1943	25.83
19872	3200	1591	13.80	1648	15.54	1703	17.30	1755	19.08	1806	20.87	1856	22.68	1904	24.51	1952	26.36	1998	28.22
21114	3400	1662	15.61	1717	17.45	1770	19.30	1821	21.18	1870	23.07	1918	24.97	1965	26.89	2010	28.83	2055	30.79
22356	3600	1734	17.58	1787	19.53	1838	21.48	1887	23.45	1935	25.44	1981	27.44	2027	29.46	2071	31.49	2114	33.54
23598	3800	1807	19.75	1858	21.79	1907	23.85	1955	25.92	2001	28.00	2046	30.10	2090	32.22	2133	34.35	2175	36.49
24840	4000	1881	22.11	1930	24.26	1978	26.41	2024	28.58	2069	30.77	2112	32.96	2155	35.17	2197	37.40	2238	39.63

C.F.M.	O.V.	6-1/2" S.P.		7" S.P.		7-1/2" S.P.		8" S.P.		9" S.P.		10" S.P.		11" S.P.		12" S.P.		13" S.P.	
		R.P.M.	B.H.P.	R.P.M.	B.H.P.	R.P.M.	B.H.P.	R.P.M.	B.H.P.	R.P.M.	B.H.P.	R.P.M.	B.H.P.	R.P.M.	B.H.P.	R.P.M.	B.H.P.	R.P.M.	B.H.P.
13662	2200	1808	19.43	1861	20.96	1914	22.53	1965	24.12	2066	27.42	2163	30.84						
14904	2400	1847	21.23	1898	22.83	1949	24.45	1998	26.11	2095	29.50	2188	33.02	2279	36.66	2368	40.42		
16146	2600	1892	23.20	1940	24.86	1989	26.56	2036	28.28	2129	31.79	2220	35.42	2307	39.16	2393	43.01	2477	46.97
17388	2800	1939	25.33	1986	27.07	2033	28.84	2079	30.63	2169	34.28	2256	38.03	2341	41.89	2423	45.84	2504	49.90
18630	3000	1989	27.63	2035	29.45	2081	31.30	2125	33.16	2212	36.96	2296	40.85	2378	44.84	2458	48.91	2537	53.09
19872	3200	2043	30.11	2088	32.01	2131	33.94	2174	35.89	2258	39.84	2340	43.88	2420	48.00	2497	52.21	2573	56.51
21114	3400	2099	32.76	2142	34.76	2184	36.77	2226	38.80	2308	42.91	2387	47.11	2464	51.38	2540	55.73	2614	60.17
22356	3600	2157	35.61	2199	37.69	2240	39.79	2280	41.91	2360	46.19	2437	50.55	2512	54.98	2585	59.48	2657	64.06
23598	3800	2217	38.65	2257	40.82	2297	43.01	2337	45.21	2414	49.67	2489	54.20	2562	58.79	2633	63.45	2703	68.18
24840	4000	2278	41.89	2318	44.15	2357	46.44	2395	48.73	2470	53.37	2549	58.07	2614	62.83	2684	67.65	2752	72.54
26082	4200	2341	45.34	2379	47.70	2417	50.08	2455	52.47	2528	57.28	2599	62.16	2668	67.10	2736	72.09	2803	77.15
27324	4400	2405	49.02	2442	51.47	2479	53.94	2516	56.42	2587	61.43	2656	66.49	2724	71.60	2790	76.77	2855	81.99
28566	4600	2470	52.92	2507	55.47	2543	58.04	2578	60.62	2648	65.81	2716	71.05	2782	76.34	2847	81.69	2910	87.09
29808	4800	2536	57.06	2572	59.71	2607	62.38	2642	65.05	2710	70.43	2776	75.86	2841	81.34	2904	86.87	2966	92.44
31050	5000	2604	61.46	2639	64.21	2673	66.97	2707	69.74	2773	75.31	2838	80.93	2901	86.59	2963	92.31	3024	98.06
32292	5200	2672	66.11	2706	68.96	2739	71.82	2772	74.68	2837	80.45	2901	86.26	2963	92.11	3023	98.01	3083	103.96

Performance shown is for air foil fan with outlet duct. BHP does not include drive loss.
Underlined figures indicate maximum static efficiency.

Courtesy of Barry Blower

Figure 4-21. Centrifugal Air Foil Fan, No. 7245, Double Width, Double Inlet.

CENTRIFUGAL AIR FOIL FAN, No. 7245

SINGLE WIDTH, SINGLE INLET

INTAKE AREA = 3.640 SQ. FT. • WHEEL DIAMETER 24-1/2"
OUTLET AREA = 3.45 SQ. FT. • TIP SPEED F.P.M. = 6.41 x R.P.M. • MAXIMUM B.H.P. = $1.785 \left(\frac{R.P.M.}{1000}\right)^3$

Top white area, Class I / Top grey area, Class II / Bottom white area, Class III / Bottom grey area, Class IV

C.F.M.	O.V.	1/4" S.P.		3/8" S.P.		1/2" S.P.		5/8" S.P.		3/4" S.P.		7/8" S.P.		1" S.P.		1-1/4" S.P.		1-1/2" S.P.	
		R.P.M.	B.H.P.	R.P.M.	B.H.P.	R.P.M.	B.H.P.	R.P.M.	B.H.P.	R.P.M.	B.H.P.	R.P.M.	B.H.P.	R.P.M.	B.H.P.	R.P.M.	B.H.P.	R.P.M.	B.H.P.
2760	800	460	0.17	505	0.23	549	0.29	591	0.36	631	0.43	670	0.50	708	0.58	781	0.75		
3105	900	495	0.21	536	0.27	576	0.34	615	0.41	652	0.49	689	0.56	725	0.65	793	0.82	858	1.00
3450	1000	532	0.26	570	0.33	607	0.40	643	0.47	677	0.55	711	0.63	745	0.72	809	0.90	871	1.09
3795	1100	570	0.31	606	0.39	640	0.46	673	0.54	705	0.62	737	0.71	768	0.80	829	0.99	887	1.18
4140	1200	610	0.38	642	0.45	674	0.54	705	0.62	735	0.71	765	0.80	794	0.89	851	1.08	907	1.29
4485	1300	650	0.45	680	0.53	710	0.62	739	0.71	768	0.80	796	0.90	823	0.99	877	1.20	929	1.41
4830	1400	690	0.53	719	0.62	747	0.71	775	0.81	802	0.91	828	1.00	854	1.11	905	1.32	954	1.54
5175	1500	731	0.62	759	0.72	785	0.82	811	0.92	837	1.02	862	1.13	886	1.23	934	1.45	982	1.68
5520	1600	773	0.73	799	0.83	824	0.94	849	1.04	873	1.15	897	1.26	920	1.37	966	1.60	1011	1.84
5865	1700	815	0.85	840	0.96	864	1.07	887	1.18	910	1.29	933	1.40	955	1.52	998	1.76	1041	2.01
6210	1800	857	0.98	881	1.09	903	1.21	926	1.33	948	1.44	969	1.56	991	1.69	1032	1.94	1073	2.19
6555	1900	899	1.12	922	1.24	944	1.37	965	1.49	986	1.61	1007	1.74	1027	1.87	1067	2.12	1107	2.39
6900	2000	942	1.28	964	1.41	985	1.54	1005	1.67	1025	1.80	1045	1.93	1065	2.06	1103	2.33	1141	2.61
7590	2200	1028	1.65	1048	1.79	1067	1.93	1086	2.07	1105	2.21	1123	2.35	1141	2.50	1177	2.79	1212	3.09
8280	2400	1114	2.09	1133	2.24	1151	2.39	1168	2.54	1186	2.69	1203	2.85	1220	3.00	1253	3.32	1285	3.64
8970	2600	1201	2.59	1218	2.76	1235	2.92	1252	3.09	1268	3.25	1284	3.42	1300	3.59	1331	3.92	1361	4.26
9660	2800	1289	3.19	1305	3.36	1320	3.54	1336	3.71	1351	3.89	1366	4.07	1381	4.25	1410	4.61	1439	4.97
10350	3000	1376	3.86	1391	4.05	1406	4.24	1421	4.43	1435	4.62	1449	4.81	1463	5.00	1491	5.38	1518	5.77
11040	3200	1464	4.63	1478	4.83	1492	5.03	1506	5.23	1520	5.44	1533	5.64	1546	5.84	1573	6.25	1598	6.66
11730	3400	1552	5.50	1566	5.71	1579	5.93	1592	6.14	1605	6.35	1617	6.57	1630	6.79	1655	7.22	1680	7.65

C.F.M.	O.V.	2" S.P.		2-1/2" S.P.		3" S.P.		3-1/2" S.P.		4" S.P.		4-1/2" S.P.		5" S.P.		5-1/2" S.P.		6" S.P.	
		R.P.M.	B.H.P.	R.P.M.	B.H.P.	R.P.M.	B.H.P.	R.P.M.	B.H.P.	R.P.M.	B.H.P.	R.P.M.	B.H.P.	R.P.M.	B.H.P.	R.P.M.	B.H.P.	R.P.M.	B.H.P.
4140	1200	1012	1.73	1112	2.21														
4485	1300	1030	1.86	1125	2.36	1216	2.88												
4830	1400	1050	2.01	1142	2.51	1229	3.05	1313	3.63										
5175	1500	1073	2.17	1161	2.69	1245	3.24	1326	3.82	1404	4.44								
5520	1600	1098	2.34	1182	2.87	1263	3.44	1341	4.04	1417	4.66	1491	5.32	1562	6.00				
5865	1700	1125	2.52	1205	3.07	1283	3.65	1359	4.26	1432	4.90	1504	5.57	1573	6.26	1640	6.98		
6210	1800	1153	2.73	1230	3.29	1305	3.89	1378	4.51	1450	5.16	1519	5.84	1586	6.55	1652	7.28	1716	8.03
6555	1900	1183	2.94	1257	3.53	1329	4.14	1400	4.78	1469	5.44	1536	6.13	1601	6.85	1665	7.59	1728	8.36
6900	2000	1214	3.18	1286	3.78	1355	4.41	1423	5.06	1490	5.74	1555	6.44	1618	7.17	1681	7.93	1742	8.71
7590	2200	1280	3.70	1346	4.33	1411	5.00	1474	5.68	1537	6.39	1598	7.13	1658	7.88	1717	8.66	1775	9.47
8280	2400	1349	4.29	1411	4.96	1471	5.66	1531	6.38	1589	7.13	1647	7.89	1703	8.68	1759	9.49	1814	10.32
8970	2600	1421	4.96	1479	5.67	1536	6.41	1592	7.17	1647	7.94	1701	8.74	1754	9.56	1807	10.40	1859	11.26
9660	2800	1495	5.71	1550	6.47	1604	7.25	1656	8.04	1708	8.85	1759	9.69	1810	10.54	1860	11.41	1909	12.30
10350	3000	1571	6.55	1623	7.35	1674	8.17	1724	9.01	1773	9.86	1821	10.73	1869	11.62	1917	12.52	1964	13.45
11040	3200	1649	7.49	1698	8.34	1746	9.19	1794	10.07	1840	10.96	1886	11.87	1932	12.80	1977	13.74	2022	14.70
11730	3400	1728	8.53	1774	9.42	1820	10.32	1866	11.24	1910	12.17	1954	13.12	1997	14.08	2040	15.06	2083	16.06
12420	3600	1807	9.67	1852	10.61	1896	11.55	1939	12.52	1982	13.49	2024	14.48	2065	15.48	2106	16.50	2147	17.53
13110	3800	1888	10.93	1931	11.91	1973	12.90	2014	13.91	2055	14.92	2095	15.95	2135	17.00	2174	18.06	2214	19.13
13800	4000	1970	12.30	2011	13.33	2051	14.37	2091	15.42	2130	16.48	2169	17.55	2207	18.64	2245	19.74	2282	20.85

C.F.M.	O.V.	6-1/2" S.P.		7" S.P.		7-1/2" S.P.		8" S.P.		9" S.P.		10" S.P.		11" S.P.		12" S.P.		13" S.P.	
		R.P.M.	B.H.P.	R.P.M.	B.H.P.	R.P.M.	B.H.P.	R.P.M.	B.H.P.	R.P.M.	B.H.P.	R.P.M.	B.H.P.	R.P.M.	B.H.P.	R.P.M.	B.H.P.	R.P.M.	B.H.P.
7590	2200	1831	10.29	1887	11.14	1942	12.01	1996	12.90	2101	14.73								
8280	2400	1868	11.17	1921	12.04	1974	12.94	2025	13.85	2126	15.73	2224	17.69	2320	19.72				
8970	2600	1910	12.14	1961	13.04	2011	13.96	2060	14.90	2157	16.84	2251	18.84	2343	20.92	2433	23.07	2521	25.28
9660	2800	1958	13.21	2006	14.14	2054	15.09	2101	16.06	2193	18.05	2284	20.10	2373	22.23	2459	24.42	2544	26.68
10350	3000	2010	14.39	2056	15.35	2101	16.33	2146	17.33	2235	19.37	2322	21.48	2407	23.66	2490	25.90	2572	28.21
11040	3200	2066	15.67	2110	16.67	2153	17.68	2196	18.70	2281	20.81	2364	22.98	2446	25.21	2526	27.51	2606	29.86
11730	3400	2125	17.07	2167	18.10	2208	19.14	2250	20.20	2331	22.37	2411	24.60	2489	26.89	2567	29.24	2643	31.65
12420	3600	2187	18.58	2227	19.64	2267	20.72	2306	21.82	2384	24.05	2461	26.34	2537	28.69	2611	31.10	2685	33.57
13110	3800	2252	20.21	2291	21.31	2329	22.43	2366	23.56	2441	25.86	2515	28.22	2588	30.63	2659	33.10	2730	35.63
13800	4000	2319	21.97	2356	23.11	2392	24.26	2429	25.43	2501	27.80	2572	30.23	2642	32.71	2711	35.24	2779	37.83
14490	4200	2388	23.87	2423	25.04	2459	26.23	2493	27.44	2563	29.88	2631	32.38	2698	34.93	2765	37.53	2831	40.18
15180	4400	2458	25.89	2493	27.11	2526	28.34	2560	29.58	2627	32.10	2693	34.67	2758	37.29	2822	39.96	2886	42.68
15870	4600	2530	28.06	2563	29.32	2596	30.59	2629	31.88	2693	34.47	2757	37.12	2820	39.81	2882	42.55	2944	45.34
16560	4800	2604	30.38	2635	31.68	2667	32.99	2699	34.32	2761	37.00	2822	39.72	2883	42.48	2943	45.30	3003	48.16
17250	5000	2678	32.85	2709	34.20	2739	35.55	2770	36.92	2830	39.68	2890	42.48	2949	45.32	3007	48.21	3065	51.14
17940	5200	2753	35.49	2783	36.87	2813	38.27	2842	39.67	2901	42.52	2959	45.40	3016	48.32	3072	51.28	3129	54.29

Performance shown is for air foil fan with outlet duct. BHP does not include drive loss.
Underlined figures indicate maximum static efficiency.

Courtesy of Barry Blower

The elements involved in fan curves are:

- CFM
- S.P.
- RPM
- BHP
- Fan RPM Curve
- Static Efficiency Curve
- Operating Point
- Design Point
- Shut-Off Point
- Free Delivery Point
- System Resistance Curves (both design and actual)

In balancing we are mainly concerned with the basic performance curve plotting CFM against static pressure. Figure 4-20 shows a typical fan performance graph with actual and design system resistance curves, fan rpm curves, horsepower curves and the CFM and SP grid.

The actual and design system resistance lines are rarely exactly the same. The resistance line shows how the CFM, SP, and HP vary at different rpms with a fixed air distribution system.

CHANGING FAN CFM'S

Let us assume, for purposes of example, that fan S-2 in the North High School case history which will be reviewed in Chapter 10, was actually about 20 percent low when it was balanced. The system was proportionately balanced and all the outlets ended up at about 80 percent of design. Now you return to the fan and want to increase the CFM from the existing 80 percent of design (7800), to the design requirement of 9,800 CFM. This is an increase over what you actually have of about 25 percent and since all outlets are proportionately balanced they will increase the same percentage. For example if an outlet was at 400 CFM it will increase to 500 CFM.

There are 3 methods in changing fan CFM's by rpm increase:

1. By using a **fan curve.**
2. **Calculations** using Fan Law No. 1.
3. Using a **combination** of some calculations and using the fan chart.

Using A Fan Curve

Using the fan curve is the simplest of all. You simply plot the actual operating point and move up the system resistance curve to the new CFM required and read off what the new SP, RPM and break horsepowers are. See Figure 4-20.

Using Fan Chart and Calculations

The second method is used if a fan curve is not available but a chart is. What you must remember here is that the static pressure always changes when the RPM is changed with a fixed fan and system. The new SP is a function of the new CFM and RPM and they must be determined first.

The first step is to calculate your new RPM for the new CFM. See the following example on the complete calculation. This is 1193 rpm. Now go to the fan chart, Figure 4-21, locate the new CFM, the 9800 CFM line and read to the right until you find the newly calculated rpm. It comes right in between 2-1/2 and 3 inches static pressure columns, which wold make it about 2-3/4 inches, and the BHP is right in between 5-1/2 and 6-1/2, making it 5 BPH.

Calculations Using Fan Law No. 1

There is a certain procedure for calculating a change in fan CFM using Fan Law No. 1.

The order to follow for calculating a fan CFM change is:

1. New RPM
2. New SP
3. Actual BHP being used
4. New BHP
5. New pulley size required at new RPM
6. New belt length needed
7. Check if new motor, wiring, starter or overloads are needed.

The following is an example of the calculations for fan S-2 North High School if it were 80 percent of design after balancing, 7800 CFM, and you wanted to increase it to 100 percent of design, 9800 CFM. Change the fan sheave and reuse the motor sheave. Current belt is 102 inches long and there is 2-inch movement available back or forth on the motor on the motor slide rail.

The existing 5 HP motor must be removed and a 7-1/2 HP one be installed. The 9-inch-diameter motor sheave must be replaced with a 7.23-inch-diameter one. The 102-inch belts can probably be reused by moving the motor an inch back.

The maximum amp draw on the motor increases from 15.2 to 22. The 12-gauge wiring must be changed to 10-gauge. The size 0 starter should be replaced with a size 1 for the 7-1/2 HP motor.

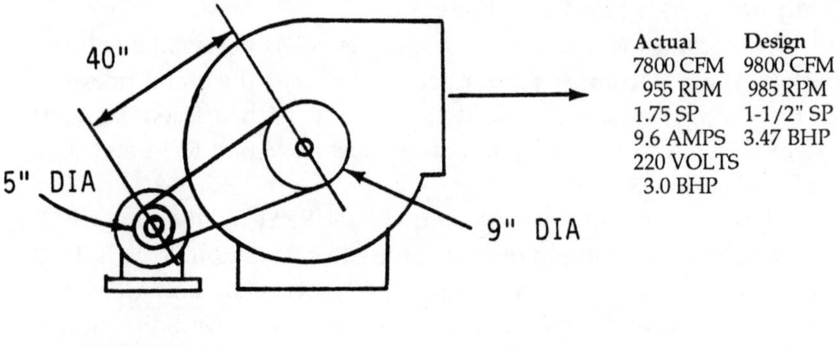

5 HP
1725 RPM
230V/3/60CY
15.2 AMPA MAX

$$\text{RPM new} = \text{RPM old}\left(\frac{\text{CFM new}}{\text{CFM old}}\right) = 955\left(\frac{9800}{7800}\right) = 1193 \text{ RPM}$$

$$\text{SP new} = \text{SP old}\left(\frac{\text{CFM new}}{\text{CFM old}}\right)^2 = 1.75\left(\frac{9800}{7800}\right)^2 = 2.7\text{" SP}$$

$$\text{BHP actual} = (\text{HP}) \times \left(\frac{\text{AMPS act}}{\text{AMPS rated}}\right) \times \left(\frac{\text{VOLTS act}}{\text{VOLTS rated}}\right) = \left(\frac{9.6}{15.2}\right) \times \left(\frac{220\text{V}}{230\text{V}}\right)$$

$$= 3.0 \text{ BHP}$$

$$\text{BHP new} = \text{BHP old} \times \left(\frac{\text{CFM new}}{\text{CFM old}}\right)^3 = 3.0 \times \left(\frac{9800}{7800}\right)^3 = 5.92 \text{ BHP}$$

$$\text{DIA fan sheave} = \text{DIA motor sheave}\left(\frac{\text{RPM motor}}{\text{RPM fan}}\right) = 5 \times \frac{1725}{1193} = 7.23\text{" DIA}$$

$$\text{BELT LENGTH} = 2C + 1.57 \times (D + d) + \frac{(D-d)^2}{4C} = 2 \times 40 + [1.57 \times (7.23 + 5)]$$

$$+ \frac{(7.23 - 5)^2}{4 \times 40}$$

$$= 80 + 19.2 + .03 = 99.23 \text{ INCHES LONG}$$

Figure 4-22. Changing fan CFM formulas.

Chapter 5
TYPES OF VAV CONTROL SYSTEMS

BASIC TYPES OF CONTROL SYSTEMS

There are various types of control systems used in HVAC systems.

1. Smaller buildings and systems generally use conventional **indirect electric controls** with low voltage wiring between VAV boxes and thermostats, etc.
2. Larger buildings and systems in the past years generally have used **indirect pneumatic controls** with a compressor.
3. **Direct digital controls** (DDC) electronic systems are being used more widely today. They can control dampers, valves, start-stop functions, sensing temperatures, flows, pressures etc. all directly. They can also modulate dampers, valves etc. smoothly with instantaneous proportionate control.
4. Electronic systems can be **hard wired** with low voltage or line voltage wiring, or they can be carrier **wireless** systems using FM transmitters and receivers.
5. A **systems powered** control system powers itself with air from the supply duct rather than using compressed air or electricity.
6. **Volumetric** type control systems for monitoring and controlling air flows in ductwork with air measuring stations.
7. **Static Pressure** type systems for monitoring and controlling pressures and air flows with pneumatic sensors.

VAV CONTROL SUB-SYSTEMS

1. **Space Temperatures–**
   A flow sensor, controller and camper in the VAV terminals along with a thermostat in the spaces vary the air volume through the VAV terminals to maintain space temperatures.
2. **Supply Fan Volume–**
   As terminals throttle, the supply duct pressure rises. This must be sensed, communicated to the fan and the fan volume reduced accordingly.
3. **Proportioning Supply and RA Volumes–**
   While maintaining building pressure and utilizing economizer cycles.
   a. One static pressure sensor in the supply duct doesn't work well because it reduces both the supply and the return fan volumes proportionately. This reduces the absolute difference between the volumes as the fan volume decreases, which means that the makeup air and exhaust air are no longer in balance and control has been lost.
   b. Using two static pressure sensors, one in the supply duct and the other in the return duct, with a reset on the return fan controller, maintains the absolute volume difference between the two fans.
   c. A very positive method for maintaining the absolute volume difference between the supply and return fans is with air flow measuring stations in the supply and return systems.

4. **Supply Air Temperature–**
   Generally supply air temperatures are maintained at a constant temperature. However, this is not necessarily the best or most economical operation. Supply air temperatures may be reset and raised with lighter cooling loads which allows savings with economizers, chilled water temperatures and possibly estimating the use of chiller completely using all outside air for cooling.
5. **Economy Cycle Control of Outside Air–**
   The economy cycle regarding mixed air control in a VAV system is the same as with a conventional HVAC system except for one important difference. The absolute volume of minimum outside in

the minimum outside air cycle decreases. When the total volume of air in a VAV system is reduced because conditions are satisfied, the minimum outside air in the VAV may be reduced proportionately.

There are two ways to rectify this situation. The simplest way, but costliest operationally, is to increase the design minimum outside air based on minimum flow conditions. The second way is to add a volume or pressure flow sensor in the outside air to maintain the outside air design minimum at all times to meet all conditions.

SUPPLY AIR SEASONAL TEMPERATURES REQUIREMENTS

The following general supply temperature should be maintained in VAV systems:

1. In **summer** the cold air temperature must be low enough for the cooling dew point.
2. In **summer** the warm air supply temperature is governed by the return air spaces.
3. In **intermediate seasons and winter** the cold air supply temperature may be raised based on outside air temperature or kept constant the year around.
4. The warm air temperature for **winter** can be readjusted from the outside air.

VOLUMETRIC CONTROLS

Air measuring stations with Pitot tube type sensors can be incorporated at various locations in the air distribution system, such as in the discharges of the supply and return fans and in the outside air intake duct. They then coordinate supply and return fan volumes and control the minimum outside air quantities and building pressure.

STATIC PRESSURE CONTROLS

The static pressure controllers monitor the changes in supply duct pressure and direct the fan volume controller to change volume accordingly.

They are more economical than air flow monitors. However, they must be located effectively so that they measure a true representation of the changing supply duct pressures. One third to two thirds of the way downstream from the supply fan generally will work well.

BASIC COMPONENTS OF VAV CONTROL SYSTEMS

The basic components used in VAV systems are as follows:

1. **Sensors** are used for sensing temperatures, pressures, humidity, flow volume and speeds.
   Thermostats, humidistats, thermometers, aquastats, pitot tubes, air flow measuring stations are typical sensors.
2. **Control devices** are devices which actually control the media in HVAC systems such as air, water, refrigerant, steam etc. Typical devices are dampers and valves.
3. **Operators** control and power the control devices. Typical operators are electrical and pneumatic control motors.
4. **Controllers, and receivers controllers (RC)** receive the information sensed, decide what is to be done by set points and programmed formulas and direct operators of controlled devices as to what to do. For example the thermometer sends message to the RC, the controller compares the temperature to a set point number and decides to pen or close a valve or damper.
5. **Transmitter.** A transmitter transmits the information derived by the sensor in a manner required by the system. For example in a DDC system air pressure sensed in psi is changed to milliamps and then transmitted to the controller.
6. **Transducer.** A transducer converts a sensed value from one set of units into the units used by the transmitter. For example, as above, air pressure is sensed with a static pressure probe and is converted into a proportionate milliamp value for transmission. A PE switch (pneumatic/electric switch) is a typical transducer.
7. **Square root extractor.** When the fan law is used to control fan volumes derived from static pressure sensing in the ductwork, a square root extractor is used. It duplicates the fan law formula for converting the static pressure into resultant fan volume.

TEMPERATURE CONTROL SYSTEM COMPONENTS
Check-Off List

Thermostats &
Temperature Sensors

__ Space Thermostat
__ Set Back Thermostat
__ Discharge Duct
__ Mixed Air
__ Outside Air
__ Return Air
__ Freezstat
__ Remote Bulb Stat

Aquastats

__ Chiller Water Discharge
__ Chiller Water Return
__ Cooling Tower In
__ Cooling Tower Out

Thermometers

__ Stem
__ Deep Well

Humidistats

Pressure Sensors

__ Pitot Tube

Valves

__ 2-Way Control
__ 3-Way Diverting
__ 3-Way Mixing
__ Pressure Reducing Valve
__ Radiator
__ Solenoid
__ Gas Valve
__ Gas Valve, Combination
__ Thermostatic Expansion Valve

Dampers

__ Opposed Black
__ Parallel Blade
__ Inlet Vane, Fans
__ Combustion Air

VAV Controls

__ Transducers
__ Transmitters
__ Static Pressure Controller

(VAV Controls Cont'd)

__ Control Panel

Panels

__ Control
__ Readout

Pneumatic Systems

__ Copper Tubing
__ Plastic Tubing
__ Compressor, Duplex
__ Pressure Control Valves
__ Drier
__ Filter
__ Motor
__ Starter

Electronic Systems

__ Relays
__ Wiring
__ Capacitors
__ Resistors
__ Timers
__ Electronic Ignition
__ Intermittent Ignition
__ Thermocouples
__ Sequencers
__ Transformers
__ Contactors
__ Pressure Controls

Control Motors

__ 2 Position
__ Modulating

Economizers

__ Outside Air/Mixed Air
__ Enthalpy Control
__ Boiler

Safety Switches

__ High Temperature Limit
__ Low Temperature Limit
__ High Pressure Limit
__ Low Pressure Limit
__ Fan and Limit Control, Furnace
__ Air Flow Switch
__ Sail Switch

CONTROL SYSTEM SYMBOLS

CONTROL SYSTEM SYMBOLS

- [T] Duct Thermostat or Temperature Sensor
- (T) Room Thermostat or Temperature Sensor
- [H] Duct Humidistat or Humidity Sensor
- (H) Room Humidistat or Humidity Sensor
- [R] Relay (Function as indicated on Diagram)
- [RC] Receiver-Controller
- [DM] Damper Motor (Operator)
- [F] Flow or Velocity Sensor
- [SP] Static Pressure Sensor
- [FR] Fire (High Temperature) Sensor/Switch
- [SD] Smoke Detector
- [LLT] Low Limit Thermostat
- Manual Gradual Switch
- (M) Control Air Supply Main
- (EP) Control Air Supply from EP Relay
- [EP] EP Relay
- [PE] PE Relay
- (Pneumatic) Restrictor
- [V] Two-Way Valve
- [V] Three-Way Valve
- Solenoid Valve
- Thermal Expansion Valve (Refrigerant)

PNEUMATIC SYSTEMS

Traditionally the conventional pneumatic approach to controlling space temperatures has been with proportional-only pneumatic thermostats acting on the VAV terminal which controls temperatures in a plus or minus 2°F range along with a term drift of 2°F per year.

Optimization of the pneumatic control system is limited in terms of changing set points, controlling wide temperature swings and performance monitoring.

Maximum and minimum flows are mechanically set. Time-consuming calibration and balancing are needed. There is no direct connection to supply volume control. Overall the operation of the conventional VAV control system is more complicated and requires continued attention.

PNEUMATIC VAV TERMINAL CONTROLS

A typical control pressure range on a pneumatic VAV box motor is from 8 to 13 psi. The 8 psi is for the maximum flow through from the box and the 13 psi for the minimum flow.

A pneumatic-electric (PE) switch energizes the perimeter VAV reheat box at around 14 psi. At no time can both the heating and cooling be on with these types of pressure settings.

PNEUMATIC THERMOSTATS

Pneumatic thermostats can be direct acting or reverse acting. Direct acting thermostats turn the control power on, and indirect acting work in reverse and turn the control power off.

DIRECT DIGITAL CONTROLS (DDC) FOR VAV

The latest development in HVAC systems controls utilizes digital microprocessors to perform controlling actions. DDC offers lower capital investment and substantial performance improvements on most medium and large size projects.

The DDC VAV control system employs a microprocessor with low voltage power supply (24V), sensors for the space temperature supply air temperature and air velocity or static pressure sensors.

The microprocessor collects data such as space temperatures, supply air temperatures, air velocities and pressures, VAV damper position, etc. The computer program through formula and input values then controls on a feed forward and backward basis for space temperatures, maximum and minimum supply air flows, and monitors and adjusts itself accordingly.

Advantages of the DDC approach are lower capital and operating costs, greater operating flexibility, reduced maintenance and self diagnostics.

ENERGY MANAGEMENT SYSTEMS

Definitions

1. **Analog Inputs**
   Monitors variables such as temperature, humidity and pressure.
2. **Analog Outputs**
   Drives, valves, dampers and motors.
3. **Digital Inputs**
   On/Off type readings for:
   - flow/no flow
   - on/off alarms
   - high/low limit alarms
4. **Digital Outputs**
   Turns equipment on/off
   Overrides for air handling units, fans, etc.

DDC VAV MANAGEMENT SYSTEMS

Centralized Energy Management Systems

A centralized system is one in which all processing and data storage takes place in a **single computer unit**. This computer interfaces to sensors, local loop controllers and display devices through multiplexing or remote field processing units. The remote field processing units are salved to a host central processing unit and have no stand-alone capability. If a remote processor is capable of performing its local functions independent of other processors, it is considered a stand-alone.

The central processing unit maintains a fully **replicated database**, while each field processing unit typically maintains a **local point database**. Changes in field point conditions are directed to the host central processing unit via change-of-state messages. The **field processing units** can execute field commands, such as resetting the setpoint of a local loop controller, when directed to do so by the host central processing unit.

A centralized system, by having all function software and data at one location, allows for rapid and integrated action. Thus, for instance, demand limiting can decide if it has a higher priority than the interlocking feature by simply examining the common point database. A disadvantage of this structure is that all capability ceases upon failure of the central processing unit unless a redundant processor is included.

DISTRIBUTED ENERGY MANAGEMENT SYSTEM, TYPE I

In a distributed processing system with centralized data, the central processing unit performs global system functions such as demand limiting, global interlocks, log and report generation. **Stand-alone processors** provide localized monitoring and control functions which do not require centralized coordination such as duty cycling, optimal start/stop, local interlocking, time programming, direct digital control. User access to local data or program software at the remote processor generally is provided through a keypad or portable terminal.

The point capacity and functionality of the stand-alone often is matched to functional or area subsystems within a facility, such as an air handling unit or a mechanical equipment room.

The stand-alone processor maintains a local database that contains interlock tables, time programmer, etc. The host central processing unit maintains a fully replicated, centralized database. Using a replicated database allows for lower-speed communications technologies such as twisted-pair and digital PBXs (private branch exchanges for telephone networks).

Stand-alone processors may communicate with the central processing unit, but not with each other. However, the processors may request global information such as outdoor temperature from the central processing unit.

Distribution system functions help avoid communication peak problems whereas centralized systems can become bogged down with data during some circumstances.

Equipment duty cycling takes place in the remote processor after receiving a duty cycle schedule from the central processing unit.

Energy management strategies such as **optimal start/stop** and **supply air reset** are local in nature, but may need data that are not local to the particular processor.

DISTRIBUTED ENERGY MANAGEMENT SYSTEM

Stand-Alone with Integrated Central Control

The second type of distribution energy management system with distributed data consists of an **optional host central processing** unit interconnected to stand-alone remote processors via a common communications bus where any node can "talk" to any other node. This scheme also supports a general broadcast capability, which is the ability of a processor to communicate simultaneously with all other nodes on the network.

The central processing unit maintains a directory of the field points. However, complete field point data such as global data and database down load information are not stored.

Implementation of the optimal start/stop feature is used as processors in the distributed network.

In the distributed database architecture, the remote processor needing data "broadcasts" a request for the information. The remote node that monitors the needed field point data intercepts the request and formats a reply message with the current point data.

DIRECT DIGITAL
ENERGY MANAGEMENT SYSTEM

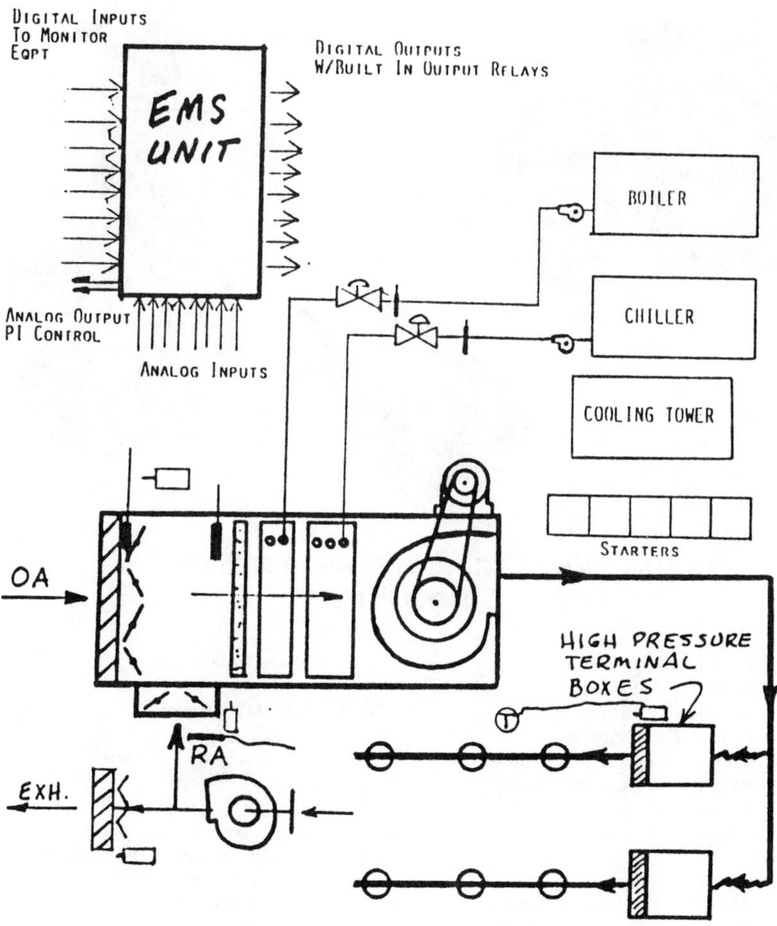

Figure 5-1

D.D.C. COMPONENTS

DDC VAV Control System Components Check-Off List

Basic EMS Equipment

__ Central EMS System
__ Modular EMS Units
__ Central EMS Systems
__ Remote Computer for Monitoring, Programming, Communicating
__ Analog
__ D.D.C.

Dedicated Controllers

__ Boiler Oxygen Trimer Controllers
__ Boiler Reset Controller, Optimizer
__ Chiller Controllers, Optimizers
__ Optimizers
__ Discharge Air Controller
__ Outdoor Reset Control System
__ Timers
__ Set Back Thermostats

Transmission Method

__ Low Voltage Hardwire
__ Powerline Carrier
__ Telephone Line
__ Multiplex
__ Remote Time Sharing

Functions, HVAC, Electrical

__ Scheduling
__ Optimum Start/Stop Time
__ Demand Limiting
__ Cycling
__ Economizer
__ Space Temperature Optimization
__ Temperature Resets, HVAC Systems
__ Chiller Optimization
__ Boiler Optimization
__ Free Cooling
__ VAV Control
__ Enthalpy Optimization
__ Fuel Changeover Sequences
__ Power Factor Correction
__ Maintenance Scheduling
__ Low Efficiency Alarming

Facilities Functions

__ Fire Control
__ Security Control
__ Building Communications

__ Process, Manufacturing Control

Controls/Accessories

__ Transducers

__ Sensors

__ Transmitters

__ Control Panels

__ Motors

VAV POINT LIST BY HVAC FUNCTION

	FUNCTION	VALUE
VAV Terminal		
Zone Temperature Sensor	AI	Zone Temperature
Zone Thermostat Reset	AO or DO	On/Off or Analog
Zone Velocity or Diff. Static Pressure	AI	Ft/Sec or Inches of Water
Zone Fan (Fan Powered Box)	DO	On/Off
Zone Reheat (Elec, HW or Steam)	AO or DO	Valve or On/Off
Supply Duct and Fan		
Supply Static Duct Pressure	AI	Inches Water Gauge
Supply Flow Rate Station	AI	Ft/Sec
Supply Duct Temperature	AI	°F
Supply Fan Pressure Difference	DI	On/Off
Supply Fan Motor Starter	DO	On/Off
Supply Fan Vane Control	AO	Vane Position
Heating and Cooling Coils		
Cooling Coil Valve	AO	Valve Position
Heating Coil Valve	AO	Valve Position
Mixed Air Low Temp. Limit	AI	On/Off (Freeze Stat)
Filter Pressure Difference	DI	Clean/Dirty
Economizer Cycle		
Mixed Air Temperature	AI	°F
Outside Air Temperature	AI	°F
Economizer Damper Control	AO	Damper Position
Building Pressurization		
Return Flow Rate Station	AI	Feet/Sec
Return Fan Pressure Difference	DI	On/Off
Return Fan Motor Starter	DO	On/Off
Return Fan Motor Control	AO	Vane Position
Enthalpy Switch Over		
Return Air Heat Content		
R.A. Dry Bulb Temp	AI	°F
R.A. RH or Dew Point Temp	AI	% RH or °F
Outdoor Air Heat Content		
O.A. Dry Bulb Temp	AI	°F
O.A. RH or Dew Point Temp	AI	% RH or °F

RETURN AIR FAN TRACKING

The control diagram below shows a VAV system with both a supply and return fan and the controls for return air fan tracking.

The controls for the supply fan include a static pressure sensor (SP-1) and flow measuring station (F-1) in the supply duct, a receiver/controller (RC-3) and damper motor (DM-4) to control the fan.

The return fan has a flow measuring station in the discharge duct and a receiver/controller (RC-4) and damper control motor (DM-5) to control the fan.

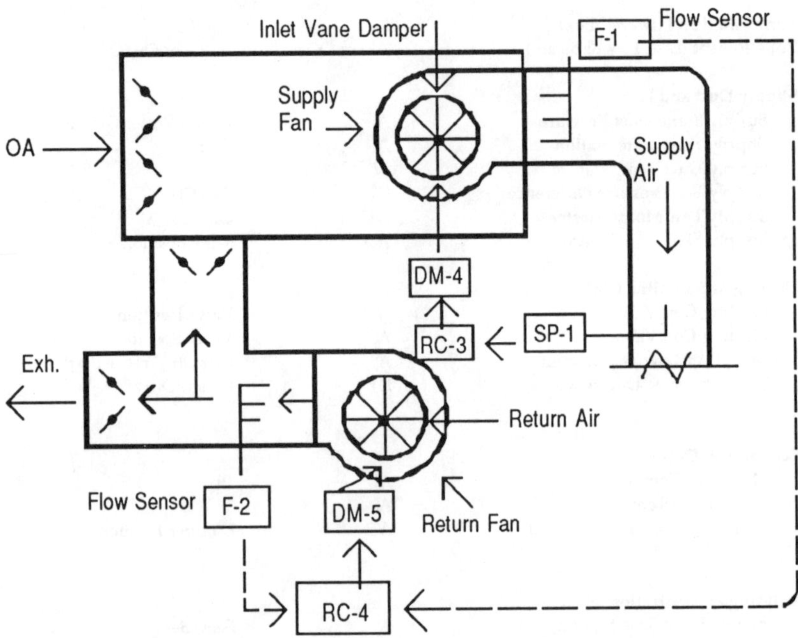

Figure 5-3. As the VAV terminals throttle down and increase the supply duct pressure, the pressure change is sensed by the F-1 static pressure sensor and is transmitted to the RC-4 controller which directs the supply fan to reduce volume.

The supply flow measuring station F-1 senses the reduced supply air flow and transmits the data to the return fan controller RC-4 which in turn directs the return damper motor to reduce flow. This is monitored by the return fan air flow station and adjusted accordingly.

If the amount of air reduced is equally in both the supply and return fans, for example 4,000 CFM in each instead of a proportional percentage decrease, the fixed amount of minimum outside air and direct exhaust are maintained as well as the resultant building pressure.

For example if a 10,000 CFM supply fan is reduced 4,000 to 6,000 CFM and the 8,000 CFM return fan reduced 4,000 to 4,000 CFM, the 2,000 CFM fixed exhaust and minimum outside are maintained.

AUTOMATIC CONTROL OF MIXED AND SUPPLY AIR TEMPERATURES RESETS

A DDC management system can automatically control **mixed air temperature** and **discharge air temperature resets** for reducing recooling and reheating energy waste.

Normally, mixed air temperature is controlled during the heating season by mixing outdoor air, at 60°F or lower, and recirculated air to maintain a constant temperature. The mixed air temperature generally ranges somewhere between 50 and 60 degrees.

The purpose in having a mixed air temperature this low in the heating season is to meet the cooling needs of those spaces requiring cooling the year around. Such spaces might be ones which have relatively high sun loads or interior zones with lighting, people and equipment loads.

When the mixed air is kept at a constant temperature, it must be kept low enough to meet the maximum cooling load ever expected in the space requiring the greatest amount of cooling. This temperature is maintained regardless of whether the spaces served actually need it or not. At times when the spaces do not need cooling the system is wasting energy because excess heating energy must be used to overcome the excess cooling effect being introduced by the outside air.

MINIMIZING MIXED AIR ENERGY WASTE

Thus, the mixed air temperature should optimally be reset to match the actual loads in the various spaces. This is done by sensing the spaces that require the greatest amount of cooling, at any given time, and resetting the mixed air temperature upwards until the needs of the space or zone are just being met.

With multizone systems, sensing the zone requirements is easily done because the positions of the hot and cold discharge deck dampers are indication of supply temperature required.

MINIMIZING COLD SUPPLY AIR ENERGY WASTE

The concept behind **cold deck discharge air temperature reset** is similar to that of the mixed air temperature reset. Match the amount of cooling supplied to the amount actually required, by resetting the cold deck discharge temperature upward whenever possible.

PRESSURE DEPENDENT FAN TERMINAL SYSTEMS

No Fan Controls
Building Pressures, Outside Air, Economizers

1. Buildings are generally best kept under an approximate 2 to 5% positive pressure. That means the air flow into the spaces should exceed the total of the return and direct exhaust air flows by roughly 2 to 5%, or the inside air pressure itself should exceed the outside pressure by .05 to .2 inches water gauge.
2. In **no economizer** is used, the outside air damper on the air conditioning unit can be manual type and be set to bring in sufficient air to cover the direct exhausts, return air and the excess supply air for positive pressure on a fixed basis.
3. If **economizers** are used, there can be a problem in maintaining the proper amount of positive pressure inside the building since the economizer allows more outside air into the building when outside temperatures are such that free cooling can be used. This consequently increases the building pressure.

   In **non-ducted return ceiling return systems** there is normally no positive way to get rid of the extra outside air and excessive space and building pressure may develop. When the bypass air is dumped into the plenum ceilings a back pressure through relief ceiling grilles may develop if the ceiling plenum pressure exceeds that of the spaces.

MAINTAINING BUILDING PRESSURIZATION

Pressure Dependent Terminals, No Fan Flow Controls

Building pressure problems can be handled in a number of ways, with barometric dampers, power returns or motor controlled dampers.

1. **Fixed Outside Air, No Power Return**

With a fixed outside air intake and no power return, if all the damper terminals are at max flows, there is no problem with meeting the design building pressurization.

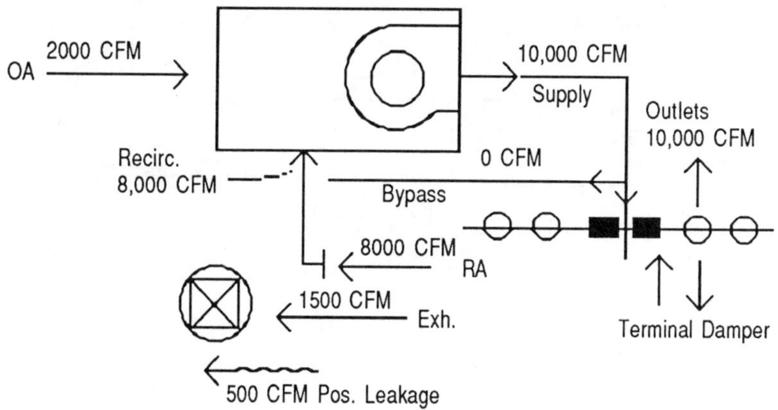

Figure 5-4.

However, if some or all the damper terminals go to maximum flow and the return air taken from the spaces doesn't reduce proportionately, the spaces can become over or under pressurized. Hence, the proper amount of bypass air and return air must be maintained. (See Figure 5-5.)

2. **Economizer System, No Power Return**

Relief air to the outside, equal to the additional outside air brought in, must be provided if an economizer is added to a damper

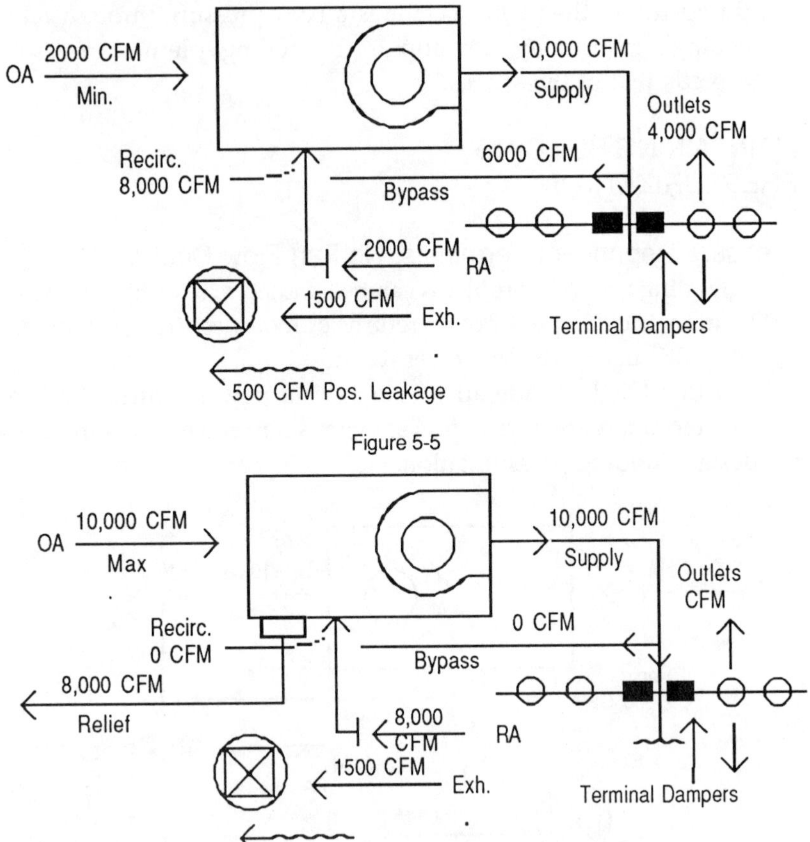

Figure 5-5

Figure 5-6

system with a fixed supply fan. At maximum supply flow and 100% outside air, the following typical conditions must be maintained. (Fig. 5-6)

At an average 70% supply air flow to the spaces and using the economizer air for cooling the outside air, recirculation and relief amounts will vary in ranges. (See Figure 5-7.)

3. **Fixed Outside Air with Power Return**

Minimum outside air is being brought in and all the damper terminals are at maximum flow. (See Figure 5-8.)

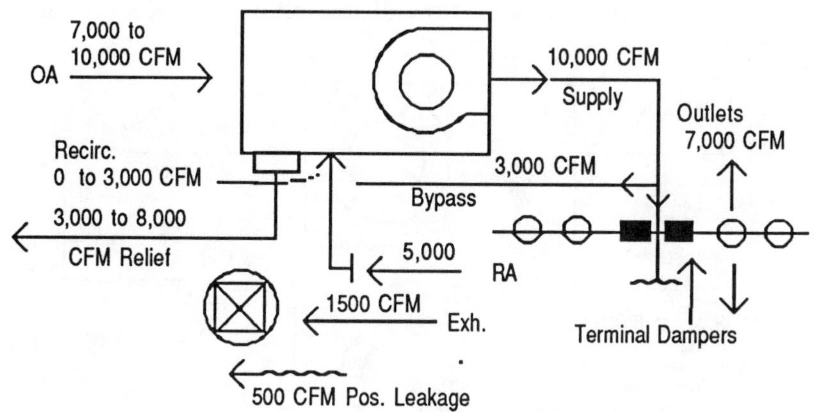

Figure 5-7.
(OA: Outside Air, RA: Return Air)

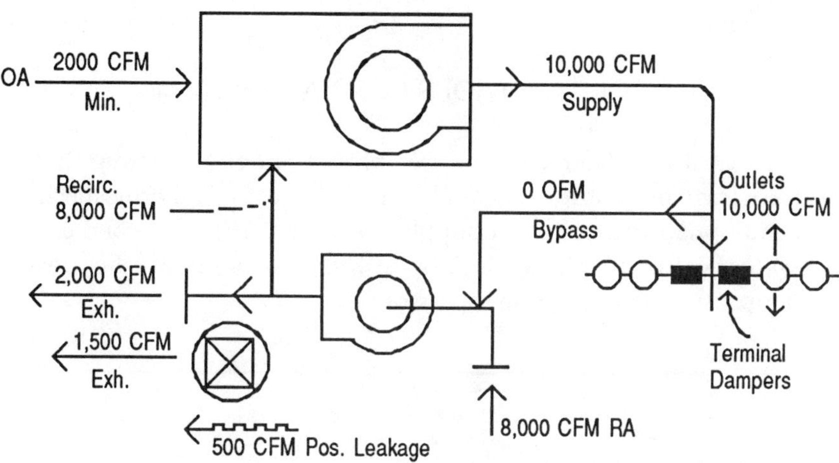

Figure 5-8.

ECONOMIZER WITH POWER RETURN

Positive exhaust air from the return air fan, equal to the additional OA brought in, must be provided if an economizer is added to the damper terminal system with a fixed supply fan.

At 70 percent average flow of the following typical conditions can exist:

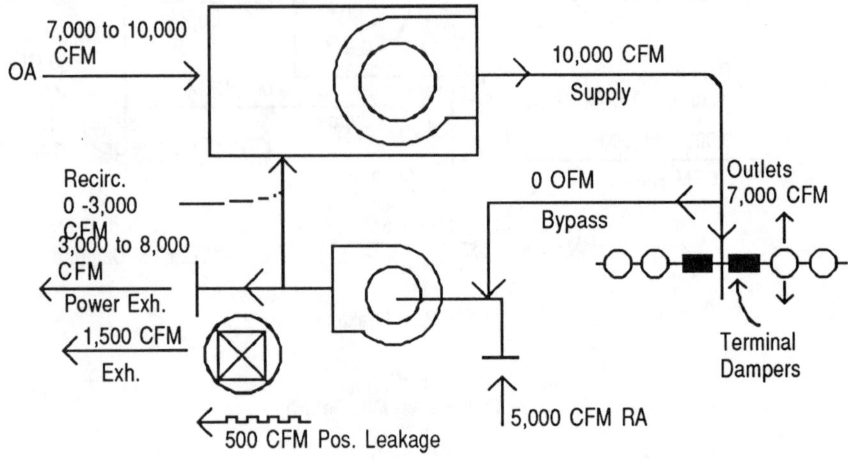

Figure 5-9.

COIL CONTROLS FOR VAV SYSTEMS

Chilled Water Control – As chilled water circulating through a cooling coil it absorbs heat from the airstream. A thermostat in the conditioned space or in the supply duct (for VAV), opens and closes a water value in the coil supply line to maintain the desired temperature in conditioned space or duct.

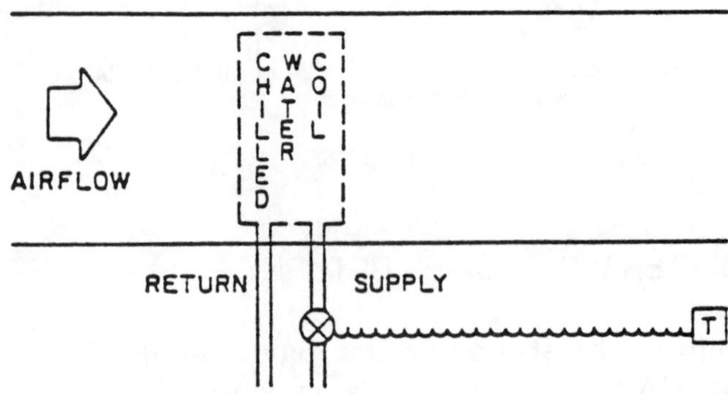

Figure 5-10.

In the diagram, a 2-way valve is shown. A 3-way valve may also be used to bypass the chilled water around the coil rather than shutting off the supply.

Direct Expansion (DX) Coil Control

Single Circuit – The DX coil can be controlled in a manner similar to the chilled water coil, using either a duct-mounted or room thermostat. In response to demand, the thermostat cycles the compressor or the condensing unit.

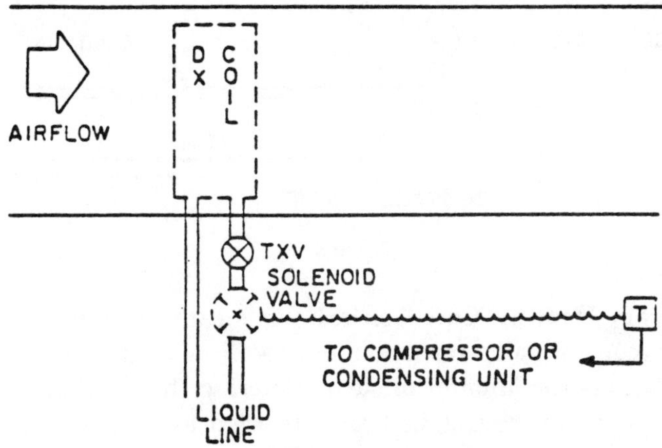

Figure 5-11.

An alternate means of control is to cycle a refrigerant solenoid valve in response to a room thermostat. This is a common method when a single condensing unit services more than one DX coil. In such cases, the condensing unit must be suitable for multi-circuit operation.

Direct Expansion Coil Control (Two-Circuit) – If a single coil has two separate circuits, the circuits can be set up to respond simultaneously or independently to demand changes.

In this example, a row-split coil is being used with a two-stage cooling thermostat. The first stage of required cooling energizes the condensing unit and supplies liquid refrigerant to coil circuit No. 1.

As cooling demand increases, the second stage of the thermostat opens the solenoid valve, allowing refrigerant flow to coil circuit No. 2 and increasing the effective cooling capacity of the coil (5-12).

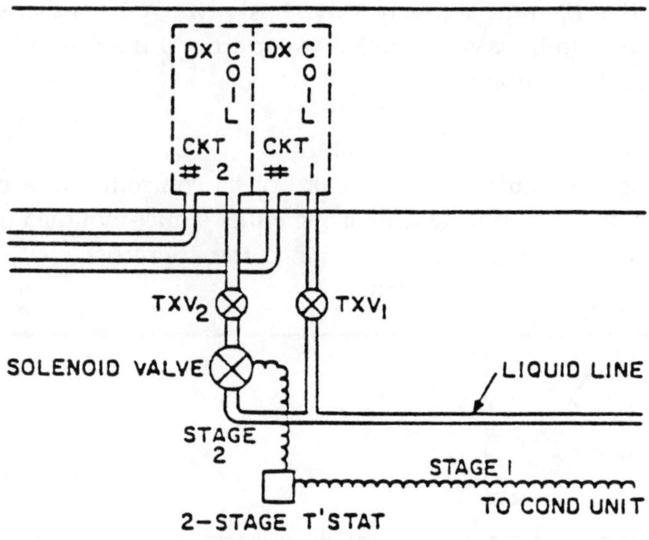

Figure 5-12.

Steam Heating Coil Control – In steam heating coils, a modulating control valve responds to a room or duct thermostat. The valve controls the amount of steam entering the coil. As the steam condenses within the coil, its heat is rejected to the airstream.

When airstream temperatures are below freezing, a double tube (IDT) coil of the nonfreezing type should be used. Where airstream temperatures are above freezing, a standard U-bend coil offers satisfactory performance and slightly lower first costs.

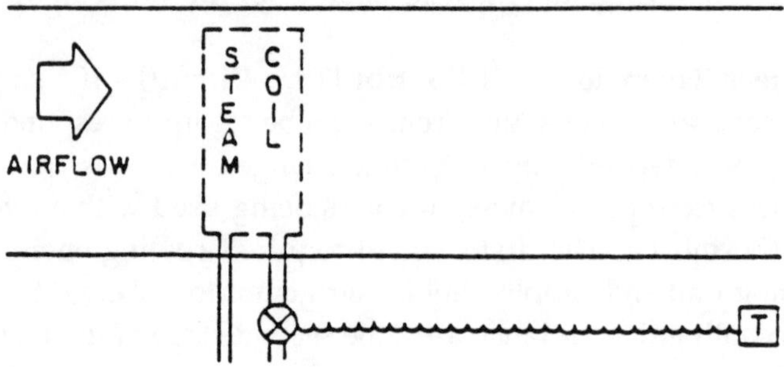

Figure 5-13.

NOTE: In preheat applications, steam should not be modulated when outside air temperature is below freezing. When the valve is closed, water trapped in the coil may freeze. In these cases, a face and bypass damper control may be desirable.

Hot Water Heating Coil Control – The hot water coil is controlled in a manner similar to the chilled water coil. In the diagram, a two-way valve is shown. A three-way valve may also be used.

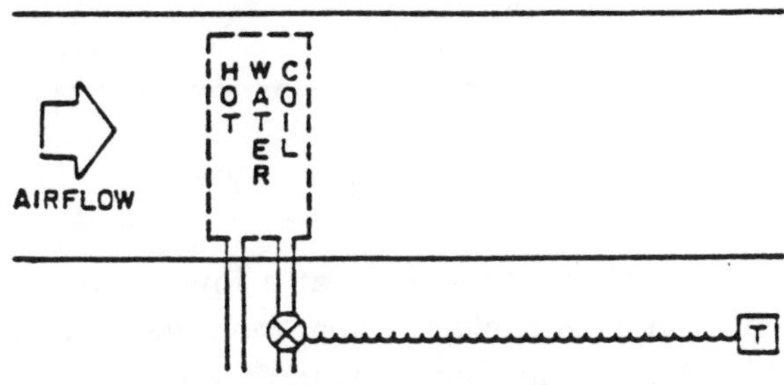

Figure 5-14.

NOTE: In preheat applications, water should not be modulated when the outside air is below freezing.

Face and Bypass Control – A set of face and bypass dampers may be used to control temperature. A thermostat in the occupied area controls a damper motor which, in turn, alters the damper blade positions to direct air through the coil for heating or, when heating requirements are satisfied, to bypass the coil.

A face and bypass control may also be used in cooling applications where dehumidification is a prime requirement.

The cooling coil is maintained at a lower temperature than that required for normal cooling so that more moisture is removed. Because the airstream is modulated with dampers the flow of water or refrigerant through the coil need not be interrupted. As a result, the coil continues to remove moisture from whatever portion of the air passes through it.

The amount of bypass air is determined by the thermostat in response to cooling demand. (See Fig. 5-15)

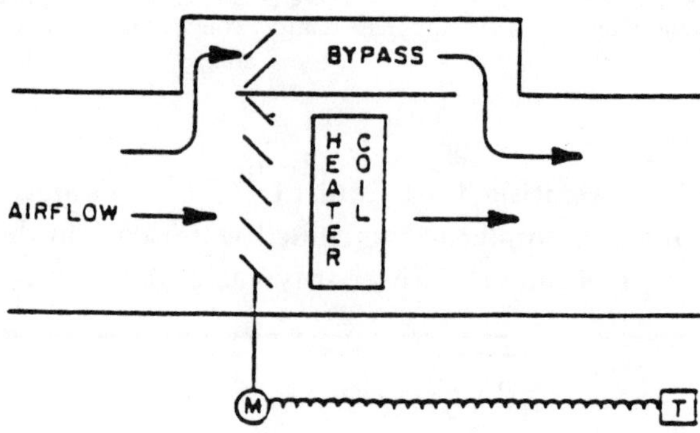

Figure 5-15.

The cooling coil is maintained at a lower temperature than that required for normal cooling so that more moisture is removed. Because the airstream is modulated with dampers the flow of water or refrigerant through the coil need not be interrupted. As a result, the coil continues to remove moisture from whatever portion of the air passes through it.

The amount of bypass air is determined by the thermostat in response to cooling demand.

Electric Heater Control – The electric heating coil consists of many electric resistance wires which generate heat when energized by an electric power source.

The resistance wires are divided into groups, or steps, that can be energized individually. This allows the degree of heating to be controlled in distinct increments.

Normally a control box is furnished with the electric heating coil. A magnetic contractor is prewired to each group of resistance wires forming a step. In this way, each step may be energized and de-energized separately. (See Fig. 5-16)

When selecting electric heat, it is very important that the face velocity through the heaters is above the minimum specified in the Electric Heating Coil Rating tables. Failure to keep the face velocity above the minimum can result in burned-out heaters.

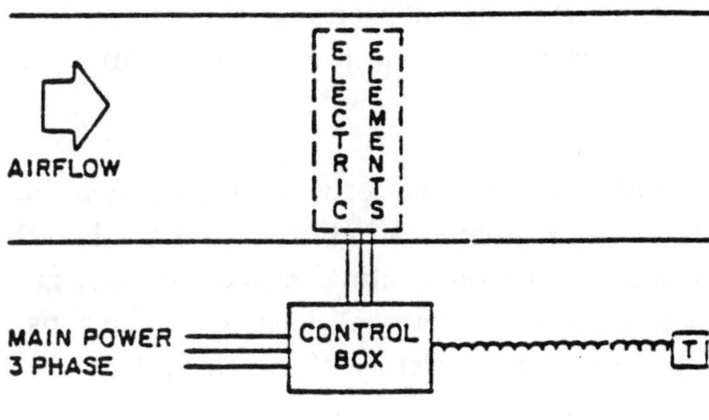

Figure 5-16.

MORNING WARM-UP AND NIGHT HEATING

VAV units with a minimum open position can be put through a night heating or warm-up cycle without a thermostat reversal by simply supplying intermittent warm air at night or for an adequate period prior to occupancy. Since there is no zone control during these periods, enough time must be allowed after the system goes back to cooling in the morning to permit cooling down any zone which may have been overheated, or an early morning buffer period must be provided.

Full Shutoff Terminals

Full shutoff terminals used in conjunction with either perimeter radiation or perimeter transmission air heating systems need no special provisions for variable volume terminals in the perimeter, since night setback or warm-up is a function performed by the perimeter system independent of the variable volume terminal. The interior system VAV terminals will be shut as long as room temperatures are below night setting, and the VAV fan may be kept idle at night.

Interior Areas

Except for terminal unit leakage, *bona fide* interior areas generally require no unoccupied heating with either type of perimeter heating. Also warm-up may be accomplished during the equalization period when the interior system is run, as a result of the wiping of the interior ceiling plenum with the warm perimeter air.

Roof Exposed Areas

Roof exposed interior areas derive no wiping heat during unoccupied periods with perimeter radiation systems, when the fans are idle, but such areas need some heating even during normal operation. For night fan economy, it is preferable to keep the fans off and to provide such heating independently of the air system.

POSITIVE WARM-UP AND NIGHT HEATING METHODS

For more positive means of warm-up or night heating, several procedures may be followed:

1. Some variable volume terminals provide a means of reversing the control sequence to permit full volumes of warm air to the space when heating is automatically or manually provided at the central equipment.
2. Dual action automatic changeover thermostat may be used for the room thermostat and control supply pressure changed in conjunction with heating provided at the central equipment.
3. Normally open variable volume terminals may be installed with a common main air pneumatic or electrical control line, deactivated whenever it is above room temperature, to throw the terminal wide open.

WARM-UP CYCLE FOR INTERIOR COOLING-ONLY VAV SYSTEMS:

1. The central heating coil turns on.
2. The cooling coil is off.
3. The return air from the perimeter heating is sensed and draws in.
4. The controllers go to reverse cooling at about 20 psi, overriding the thermostat, meaning that more air will be called for as the

temperatures in the spaces drop rather than less as is the case in the direct cooling mode.

5. Warm air then flows through the VAV boxes, rather than the normal cool air.

ENERGY MANAGEMENT SYSTEM COSTS (Indirect, Low Voltage, Hardwire)

Equipment Prices

10 Channel EMS unit W7010H*......................$3110
20 Channel EMS unit W7020H........................$5100

Installed Prices

10 Channel @ $1200/pt..............................$12,000
20 Channel @ $1000/pt..............................$20,000

Communications Link Hardware

Remote Computer for programming and monitoring
....$2000 to $3000
(Can put whole program on a floppy disk)

Functions

- Time of Day scheduling
- Optimum Start/Stop
- Demand Limiting
- Duty Cycling, Temperature Compensated by Temperatures in Spaces
- Monitoring Outside Air and Indoor Air Temperatures

**Approximate Honeywell Contractor Price*

BUDGET ESTIMATING CONTROLS FOR BUILT-UP SYSTEMS IN NEW BUILDINGS

	Percent of Equipment Costs
Under $25,000	24%
$25,000 TO $50,000	20
$50,000 TO $100,000	17
$100,000 TO $250,000	15
$250,000 TO $500,000	12

Add for computerized controls + 10%

CONTROLS FOR PACKAGE SYSTEMS

Under $25,000	8%
$25,000 to $100,000	7
$100,000 to $500,000	6

PNEUMATIC TEMPERATURE CONTROL COMPRESSOR SYSTEMS

Includes Compressor, Tank, Tubing, Dryer, Filter, PRV Valve, Motor, Starter

HP	Direct Material Cost				Total Material & Labor	
	Duplex Compressor	Other Items	Total Mat'l Costs	Man Hours	Direct Cost	With 30% O&P
1/2	$909	$3,575	$4,484	20	$5,124	$6,661
3/4	1,062	4,720	5,782	21	6,454	8,390
1	1,221	4,856	6,077	23	6,813	8,857
1-1/2	1,496	5,230	6,726	25	7,526	9,784
2	1,729	5,941	7,670	30	8,630	11,219
3	2,002	6,730	8,732	35	9,852	12,808
5	2,244	8,376	10,620	40	11,900	15,470

1. Includes installation of compressor, piping hook-up, valves, etc.
2. To furnish and install 3/8"-diameter copper trunk line tubing add $6.25 per foot.
3. In poly-vinyl tubing is used instead of copper tubing deduct 25% from copper tubing.

Direct labor costs are $32.00 per hour.

TWO-WAY CONTROL VALVES

DIA.	CV	DIRECT COST	TYPE CONNECT & PRESS	LABOR	TOTAL MATERIAL & LABOR	
Inches		Each		Man Hours	Direct Cost	With 30% O&P
1/2	.63, 1.0, 1.6	$164.02	Screw, hp	0.55	$182	$236
1/2	2.5, 4.0	101.48	Screw	0.55	119	155
3/4	6.3	171.10	Screw, hp	0.60	190	247
3/4	6.3	105.02	Screw	0.60	124	161
1	10.0	200.60	Screw, hp	0.65	221	288
1	10.0	112.10	Screw	0.65	133	173
1-1/4	16.0	228.92	Screw, hp	0.77	254	330
1-1/4	16.0	138.06	Screw	0.77	163	212
1-1/2	25.0	271.40	Screw, hp	0.83	298	387
1-1/2	25.0	184.08	Screw	0.83	211	274
2	40.0	250.16	Screw, hp	0.88	278	362
2	40.0	237.18	Screw	0.88	265	345
2-1/2	63.0	328.04	Screw	2.00	392	510
2-1/2	63.0	661.98	Flange	2.00	726	944
3	100.0	442.50	Screw	2.10	510	663
3	100.0	792.96	Flange	2.10	860	1,118
4	160.0	1097.40	Flange	2.40	1,174	1,526

Figure 5-17.

THREE-WAY CONTROL VALVES

MIXING

DIA. Inches	CV	DIRECT COST Each	TYPE CONNECT & PRESS	LABOR Man Hours	TOTAL MATERIAL & LABOR Direct Cost	TOTAL MATERIAL & LABOR With 30% O&P
1/2	4.0	$127.00	Screw	0.55	$143	$186
3/4	6.3	141.00	Screw	0.60	159	207
1	10.0	164.00	Screw	0.65	183	238
1-1/4	16.0	192.00	Screw	0.77	214	278
1-1/2	25.0	227.00	Screw	0.83	251	326
2	40.0	267.00	Screw	0.88	292	380
2-1/2	63.0	723.00	Flange	2.00	782	1,017
3	100.0	856.00	Flange	2.10	918	1,193
4	160.0	1432.00	Flange	2.40	1,503	1,954

DIVERTING

DIA. Inches	CV	DIRECT COST Each	TYPE CONNECT & PRESS	LABOR Man Hours	TOTAL MATERIAL & LABOR Direct Cost	TOTAL MATERIAL & LABOR With 30% O&P
2-1/2	63.0	$943.00	Flange	2.00	1,006	1,308
3	100.0	1056.00	Flange	2.10	1,124	1,461

NOT INCLUDED:
1. Control motor
2. Linkages
3. Plug in balance relays

hp: High pressure

Direct labor costs are $32.00 per hour

Figure 5-18.

Chapter 6
VARIABLE VOLUME FUME HOOD EXHAUST SYSTEMS

VAV fume hood exhaust systems generally involve high initial costs as well as high energy costs.

Variable volume fume hood exhaust systems involve VAV exhaust fans for different levels of exhaust, exhaust air sensing at the fund hood openings, exhaust air volume control dampers, VAV exhaust and supply fans, VAV supply terminals, reheat coils, room pressurization controls and a heating, cooling thermostat.

A typical fume hood continuously exhausts 1200 CFM and costs are $2500 per year for heating, cooling and fan power operating costs.

The average cost of a single hood installation is $15,000 to $30,000 which covers the hood, blower, ductwork, HVAC/make-up air system, controls, increased chiller and boiler capacity, and installation.

PROBLEMS

There are problems with the loss of fumes from the hoods and laboratory airflow imbalance.

It is **unsafe to turn off hoods** due to chemical storage in hoods, ongoing experiments, and potential for cold, contaminated air blowing back into room from roof. Hence, practical energy savings is small due to actual user compliance and the percent of time it is safe to turn off the hood. Systems require controls to reduce make-up air proportional to number of hoods turned off.

HOOD CONTROLS

Exhaust volume is varied usually automatically between a high and low value. Sash typically hits the microswitch at some low sash position to change air volume. Poor face velocity control limits the permissible reduction in flow due to high face velocities encountered for sash positions above switch position. There is a **hazard to operate** due to unsafe high face velocities.

Practical energy savings are low, based on low permissible turndown (typically 20 to 40%), systems require controls to reduce make-up air proportional to the amount of turndown and the number of hoods at low flow.

NIGHT SETBACK

The hood exhaust flow is reduced at night regardless of sash position. This risks loss of fumes from hoods due to lower face velocities for fully open hoods at night.

Also, the permissible setback is limited by concerns over loss of fumes from hoods.

There are low savings due to a small permissible setback in the area of 20 to 40%.

VAV FUME HOOD CONTROL

The VAV fume hood control maintains a constant face velocity to ensure safety and varies exhaust and make-up air volume to save energy.

The make-up/supply air is usually controlled to provide a slight negative air balance or airflow into the room.

There are two approaches to controlling the negative air balance. One is velocity sensing and the other sash position sensing.

BENEFITS OF SASH POSITION SENSING

1. **Energy Savings**
   There is a dramatic reduction in the loss of conditioned air and a 60 to 70% savings generated with an under-2-year payback.
2. **Lower HVAC Capital Costs**
   There is a reduction is required chiller, boiler and make-up air fan capacity through diversity of hood operation.
   The VAV hood systems solve chronic negative building pressure problems by reducing average exhaust volume. They reduce or eliminate the need for increased supply or make-up air capacity.
3. **Increased Lab Safety**
   There is a positive control of face velocity. There are alarm and monitoring features in the room air flow.

SASH SENSING

In the **sash position sensing approach,** constant face velocity over the entire hood opening area is maintained. Rather than utilized velocity sensors to measure and control air velocity at a particular point, this approach insures more accurate measurement by directly measuring sash position. Exhaust volume can then be varied automatically in direct proportion to the area of the entire hood opening as determined by sash height. Rapid response time will ensure no loss of fumes when the sash is raised or lowered. This results in a truly linear and precise system that will maintain any required face velocity. (See Figs. 6-1, 6-2, 6-3)

PLOT OF SASH AIR FLOW

Figure 6-4 shows a typical graph of air flow versus sash position. Two regions of operation are readily apparent: Constant

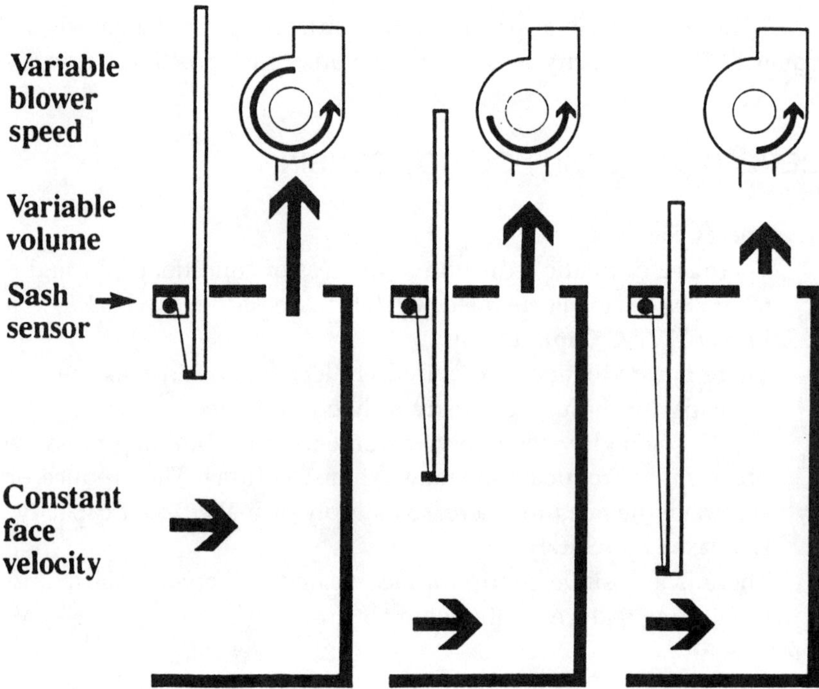

Fig. 6-1. Proportional Control of Exhaust Volume through Sash Sensing.

volume operation from 0 to 6" of sash position and variable volume from 6" to 30". A constant volume region is required due to the need for a minimum exhaust volume from the hod, 20% of full flow in this example. A minimum flow is required to purge the hood of fumes from stored chemicals, or maintain minimum lab ventilation standards. It may also provide a minimum level of exhaust to allow control of space conditions.

Once this minimum CFM and the face velocity is determined the system can be calibrated to generate the performance shown. The system is easily calibrated by setting the face velocity at both the minimum position (6") and the maximum position (30"). Three straightforward adjustments can be used to set the position of the two points on the graph with the aid of a velometer. Makeup/supply air flow can be similarly adjusted to set a constant volumetric offset between the room's total exhaust and supply volume.

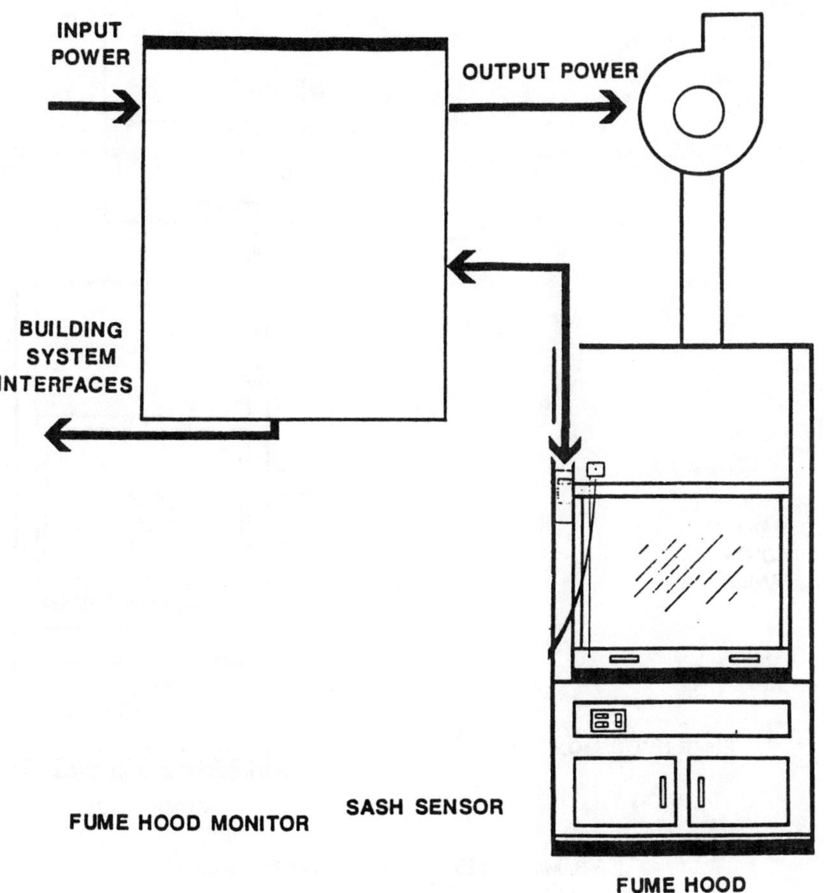

Fig. 6-2. Single Exhaust System Configuration.

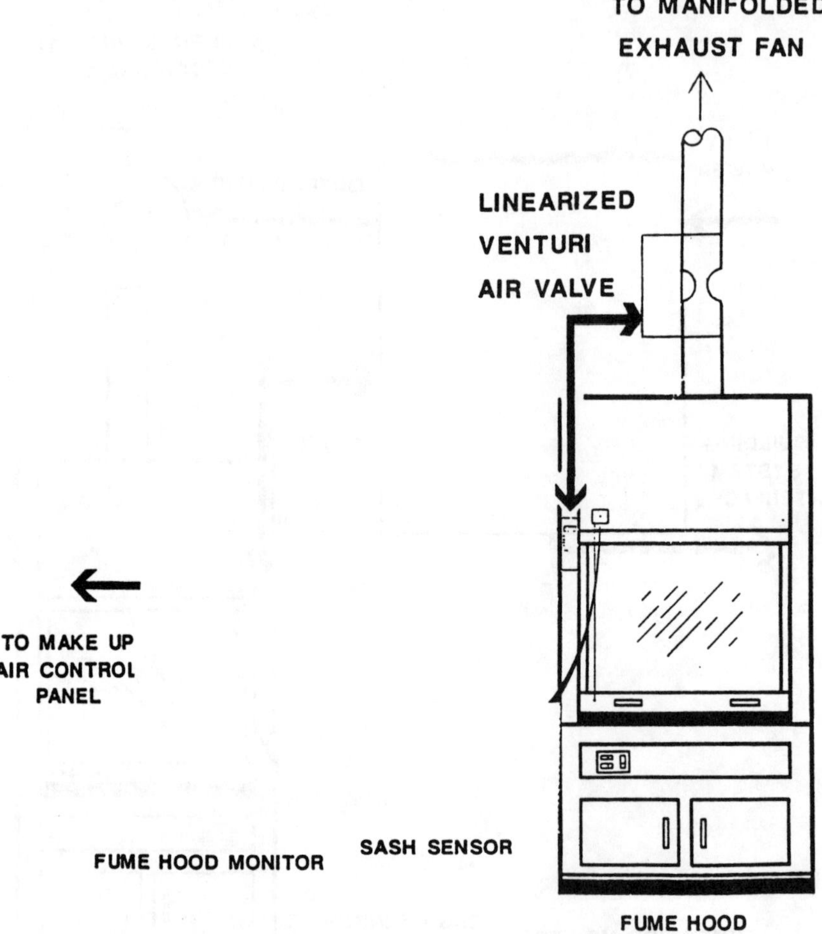

Fig. 6-3. Manifold Exhaust System Configuration.

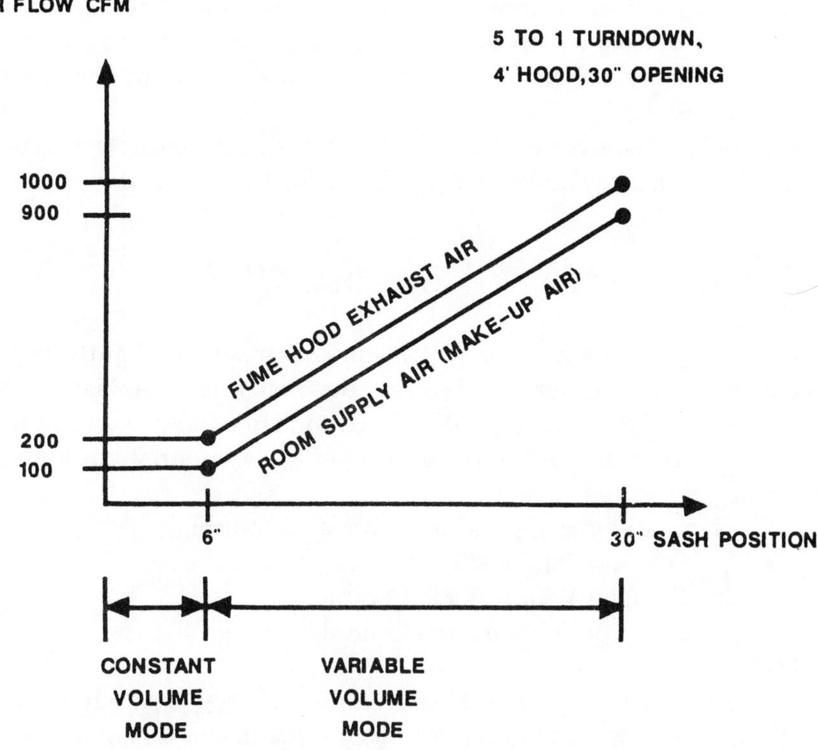

Figure 6-4. Typical Plot of Laboratory Air Flow vs. Sash Position.

DESIRED FEATURES OF VAV LAB CONTROL

1. **Remote monitoring sash position** which allows user compliance and energy savings to be monitored through linear and two state (high/low) output signals.
2. **Integrated control of make-up air** is varied to maintain a constant volume offset between a room's total exhaust and supply volumes.
3. **Integrated control of laboratory room temperature** with a general exhaust or return valve controlled to allow increased supply flow while maintaining lab pressurization.
4. **Low face velocity alarm** alerts operator to a severe reduction in face velocity.

5. **Emergency exhaust push button** overrides normal system operation for maximum exhaust in case of a lab spill or fire.
6. **Capability for remote switch** or DDC actuation of emergency exhaust mode.
7. **Digital or analog output signals** to provide remote detection of a dangerous low flow or emergency exhaust condition at the hood.

MAKEUP AIR FOR VAV HOOD EXHAUST

The goal of a makeup air system is to maintain **slightly negative pressure** to prevent any fumes in the room from escaping into the corridor. Supply air is typically varied to maintain pressurization.

The exact amount of pressurization is not important as long as net negative or positive.

There are three approaches to pressure control:

1. Constant Pressure Control
2. Constant Volume Ratio Control
3. Constant Volume offset Control

The constant volume offset control system is a simple proven approach that has been used for many years in constant volume lab rooms.

In the constant volume offset control the supply volume is controlled to maintain a fixed volume offset from the total exhaust volume. This is unaffected by open doors, loose ceiling tiles, etc.

Net pressurization is easy to maintain particularly at low exhaust levels where the net offset is large compared to the air volumes being controlled.

Building balance is straightforward to set up and maintain due to constant, set volumes drawn from the corridor.

INTEGRATED LABORATORY TEMPERATURE AND PRESSURE CONTROL

Many labs use supply air both for temperature control and as fume hood make-up air. These labs may also have a potential for cooling loads that require volumes in excess of the hood's minimum make-up air requirements. Thus supply air may at times need to be increased above the hood's exhaust volume. To maintain a net negative pressurization a general exhaust valve is typically used to exhaust additional air.

Two common strategies to control the supply and general exhaust (or return) valves consist of the following:

1. Direct thermostat control of general exhaust
2. Volumetric difference control of general exhaust

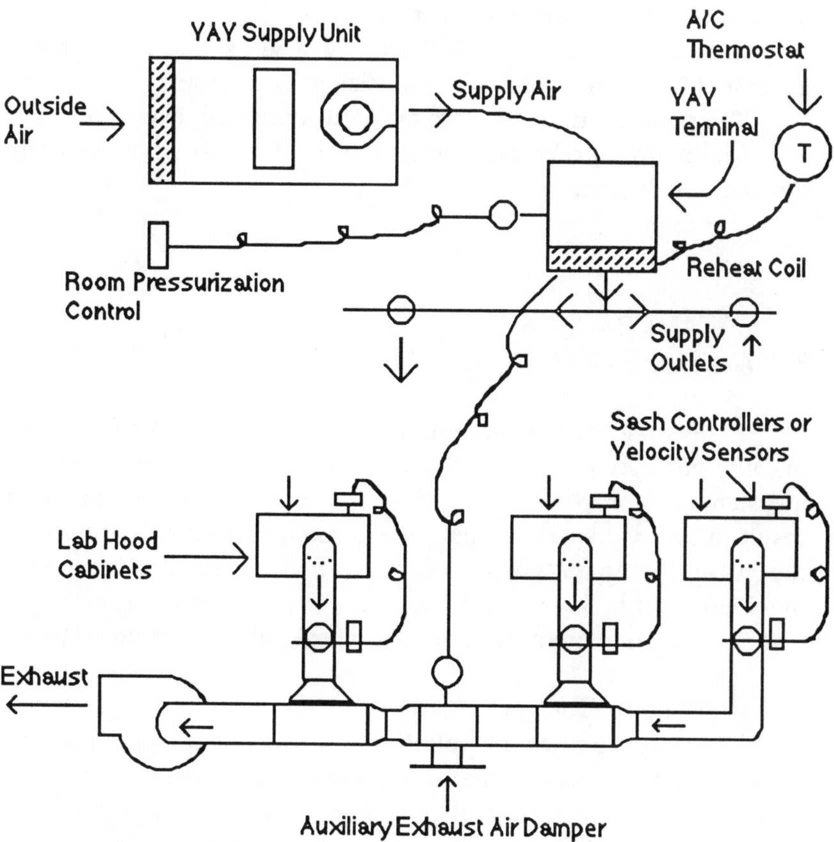

Fig. 6-5. Variable Air Volume Fume Exhaust System.

VOLUMETRIC DIFFERENCE CONTROL OF GENERAL EXHAUST

Thermostat output is appropriately scaled and high selected against hood's make-up air requirements. Resultant value controls lab's supply valve. General exhaust volume controlled by algebraic difference of supply and hood's make-up column. See Figure 6-5.

Supply is controlled by the high of lab's temperature and pressurization requirements versus the sum of the two. Result: lower supply box sizes and minimum supply air capacity.

General exhaust volume rapidly tracks the fume hood exhaust volume in a reverse acting manner.

Exhaust system's capacity is reduced via sizing for higher of general of fume hood exhaust versus sizing for their sum.

Temperature transients are avoided when sash is raised.

Room pressurization is maintained at all times even under dynamic conditions.

IMPORTANT BACKGROUND KNOWLEDGE FOR VAV EXHAUSTS

For people engaged in design and construction, VAV fume hood installations may carry with them a risk in terms of performance. If a system is improperly designed, or if attempts to keep costs down result in an inadequate system, the consequences can be costly. To design and estimate costs completely and accurately, the people involved must be knowledgeable of the basic engineering principles and consideration involved in properly designing workable systems.

The Purpose of Control

The purpose of an air pollution control system is to: 1) remove harmful contaminants such as dusts, fumes, mists, vapors, gases and heat from a source area; 2) then transport them through ducts to a collector or fan; 3) and finally separate the containants from the air and either reclaim or dispose of them, or discharge the mixture directly to the atmosphere if permissible.

Contaminants include anything that is harmful to humans, animals, plants or property in the internal or external environments. There are two basic categories of contaminants, particulates and gases.

> **Particulates** are small solid or liquid particles such as dusts, powders, smoke, liquid droplets and mists. Flash from coal furnaces and asbestos dusts are examples of harmful particulate pollutants. **Gas** pollutants are fluids without form that occupy a space rather uniformly, such as carbon monoxide or chloroform. A fume is an irritating smoke, vapor, or gas.

An important consideration in air pollution control is the harmful effects of the contaminants involved. Pollutants can be toxic (poisonous, possibly fatal) or noxious (harmful to health but not necessarily deadly). They may be corrosive (alkalis, acids, salts, and caustics) or erosive (causing wearing away by abrasion). Processes can be heat producing, moisture producing, or smoke producing, causing uncomfortable or harmful levels. Odors may be offensive or harmful. Pollutants can be inflammable, explosive, or radioactive. And finally, contaminants include bacteria and viruses.

Fire and explosion hazards are critical factors. Dusts as well as many fumes and mists have flash point temperatures, at which spontaneous combustion occurs, and explosive limits, which are percentage ranges defining when an explosion occurs.

There are two flash points for a hazardous pollutant, one for a closed situation and the other for an open situation. There are also two explosive limits, a lower and an upper. For example, ethylacetate has flash points of 24° and 30°F and lower and upper explosive limits of 2.18 and 11.4 percent. Similarly, methanol has flash points of 54° and 60°F and explosion limits of 6.72 and 36.5 percent.

With regard to toxicity, the point at which a mixture of pollutant and air starts to become toxic is called its **threshold limit value**. There are also different levels of acceptable concentrations for different lengths of exposure. Threshold limits and prolonged exposure limits are usually stated in terms of **ppm**, parts of contaminant per million parts of air, at a certain temperature and pressure. For example, ammonia has a threshold limit value of 25 ppm, a daily exposure limit of 100 ppm, and a 1/2- to 1-hour exposure limit of 2500 ppm; it is rapidly fatal at 5000 ppm.

Other measures of contaminant concentration or density sometimes used in pollution control work include **weight per cubic foot, pounds per hour, grains of moisture per pound of air, percentage of gas,** and others.

Another important factor affecting the performance of a system is the **settling rate** of solid or liquid particles. This is a measure of the time it takes for a particle to float down and come to rest, and is dependent on particle size. A 50-micron particle will free fall at 14.8 fpm, whereas a 1-micron particle will free fall at 0.007 fpm. Thus, the former will fall almost 15 feet in 1 minute; the latter will take approximately 36 hours to fall the same distance.

System Design

The general procedure in system design is to identify the problem, analyze it, establish facts, determine the magnitude of the pollution problem, calculate capacities, design hoods and ductwork, select system components (collectors, fans, makeup air units, etc.), and choose a disposal method.

Capture velocity – One of the major considerations involved in the design of an air pollution control system is the velocity required to pick up the contaminants from their source area. This velocity is called the capture velocity, and it must overcome spillage, air and heat currents, and the contaminants' weight and velocity.

For example, laboratory fumes heavier than air may require a capture velocity at their point of emission of 50 fpm to overcome their tendency to spill to the floor as well as any currents of air flowing across the source area. Sawdust, which weighs approximately 12 pounds per cubic foot, usually requires a suction pressure of 2.5 in. WG for proper pickup.

Transport velocity – The air velocity required in the duct to carry the material, overcome resistance and turns, lift it in risers, and convey it to collector or fan is the transport velocity. Powdered coal requires a transport velocity of approximately 4000 fpm.

Hood face velocity – This is the velocity required at the face of a hood for proper contaminant pickup and suction into the hood. A typical slot hood face velocity might be 2000 fpm; a kitchen exhaust hood might require a face velocity of only 100 to 150 fpm.

Hoods must be designed to draw in pollutants properly. Pollutants can be captured by completely enclosing the source area, by providing a local hood at the exact point of emission, or by providing a general exhaust hood or grille. There are an infinite number of hood configurations, and the best designs are those in which contaminants project directly into the hoods. For example, particulates from grinding, buffing, polishing, or saw wheels would be propelled directly into their hoods. For heat removal, fumes would rise into the hoods.

Air volume – Factors that determine the volume of air required are the surface area of the source, the quantity of material being removed, the amount of dilution required, the size of the hood opening, and the face velocity. A kitchen hood with a 30-square-foot opening and a 100 fpm face velocity requires 3000 CFM.

Ductwork – There are two general approaches to designing a duct system. One is the balanced static regain approach, in which the system is automatically balanced through exact pressure design of the ductwork. This can be done when all hoods are used whenever the system is in operation. The second basic approach involves blast gates. Dampers are put in branches to control air flow, and the system is not designed to be automatically balanced. Here, velocities are the main consideration. Total system static pressure is contingent on the run with the greatest resistance as well as on the pressure drops of the collector and other components.

Diversity – Some systems are designed for 100 percent usage of hoods all the time; others for only partial use of hoods at any one time. In sizing collectors and fans, the diversity factor must be taken into account to determine whether they should handle 80, 60, or some other percentage of the design air volume and static pressure. If only 60 percent of a system is in operation at any given time, it is very costly to size for larger capacities.

Material removal rate – Another factor that must be integrated into the design is the removal rate for dusts, particles, etc. The density of the material, its concentration in the air stream, the quantity of air involved, and the quantity of material to be removed must be considered to arrive at an ultimate removal rate.

AIR POLLUTION EQUIPMENT

There are four basic types of collectors, each with its own variations. These are: **centrifugal cyclones, baghouses, scrubbers, and electrostatic precipitators.** Collectors are rated by the size of particles they can remove, by percentage efficiency of removal of different size particles, and by permissible temperature ranges, pressure drops, air volumes, and water usage.

Cyclones are dry centrifugal collectors that remove particles down to approximately 10 microns at efficiencies in the 80 to 90 percent region. Particles are removed by centrifugal force and drop down into a hopper while the cleaned air goes out the top of the unit.

There are three variations of cyclones. The first is the large diameter, low resistance, type, which is most efficient in the 40- to 60-micron particle size range. The second is the medium efficiency cyclone, which has a smaller diameter housing and is generally used for 20- to 30-micron sized particles. The third cyclone is the high efficiency, small diameter model, which is very efficient in the 10- to 15-micron range.

Baghouses are dry collectors that clean like vacuum cleaners, incorporating a number of tubular, stocking-like fabric filter bags. They can collect dry particulates such as dust and fumes only and cannot be used for gases or liquids. They are highly efficient, reaching a 99.9 percent level of 1-micron size particles, and operate best in the 0.25-micron size range. Pressure losses are higher, around 3 to 8 in WG and are dependent on the air-to-cloth ratios involved.

There are two basic types of fabric collectors, distinguished by the method of bag cleaning and direction of air flow. The mechanical baghouse has motorized shakers to remove the dust from the insides of their vertical tubes. The dust-laden air flows up and into the tubes, then out through the fabric. The other type of baghouse is pneumatically cleaned–air is used to clean the bags rather than mechanical shakers–and there are three variations: **reverse air flow, reverse jet** and **pulse jet.**

Wet scrubbers can be used for particulates, gases, and liquids. Liquid droplets in a scrubber capture particulates mechanically while gases are removed through absorption. There are a number of different types of scrubbers such as packed towers, venture, wet centrifugal, wet dynamic, orifice types, and fog towers. Efficiencies range from 90 to 99 percent for particles in the 1- to 5-micron range. High energy scrubbers remove particles in the 0.25- to 1-micron range. Wet scrubbers are often divided into three groups according to pressure losses, there being 4 to 8, 8 to 25, and 25 to 60 in. WG respectively.

Electrostatic precipitators are used primarily to collect particulates such as welding fumes and some mists. Efficiencies are in the 98 to 99 percent range for 0.25- to 1-micron particles. A precipitator employs an ionization process whereby incoming particles are made negative and then collected on a positive plate through magnetic attraction. The collector plate is periodically cleaned by rapping, causing the particles to fall away by gravity.

INSTRUMENTS TO USE FOR BALANCING FUME EXHAUST SYSTEMS

For large hood openings such as kitchen and fume hoods or entrances to paint spray booths, use an anemometer and take a traverse. Divide the opening up into equal areas and take equal spot readings at the center of each area for a total of a minute. For example, if there are six equal areas, take a 10-second reading of each, going from one to the other without stopping. Use 100 percent of hood area for area factor.

Small hood opening with low velocities – Traverse either with the velometer hooked up with the low flow probe and average the readings or use the anemometer. Use a .90 correction factor on the hood area.

Slot hood with high face velocities – If the velocities are anywhere from 500 to 2000 or 3000 fpm, use the velometer with the diffuser probe. Reverse the connections for exhaust readings. Use a .85 correction on the hood opening area.

Slot hood with low face velocities – If the velocities are under 500 fmp, traverse with a thermal anemometer and average the readings.

The most accurate method of reading air flow through a hood and the method accepted by OSHA is to traverse the branch duct with a pitot tube. Pitot traverses should be made in each branch duct to a hood to determine CFM flow. In cases where a duct is not accessible, the readings can be taken at the hood.

There must be a suitable length of straight duct and velocities cannot be below 800 fpm. If branch ducts are traversed for air flow, take center line statics in the branch ducts as reference points for rebalancing, call banks, changes, etc.

SYSTEM CONDITIONS AND CAPTURE VELOCITIES

Be aware of what the capture hood face and duct transport velocities must be for the particular substance being exhausted when balancing.

Balance under actual conditions if at all possible so that observations can be made of how well contaminants are being exhausted. All that matters in the end is that the contaminant is actually captured at the source, sucked into the hood and transported to the collector.

If you cannot balance under actual conditions, assimilate hoods and capture power at the source, with smoke, a candle, steaming water pan, sand, powder, etc.

BALANCING PROCEDURE

1. **Prepare Test Reports:** Fill in equipment and system design data on the fan sheet and list hoods etc. on outlet balance sheet along with sizes, required CFM's, etc.

2. **Check Out Equipment:**
   - Check the motor name plate, starter, overloads, fan wheel, drives and bearings.
   - Bump the fan for rotation.
   - Check back draft damper at fan discharge.
   - Check the collector. Are the filters or cleaning media in? Are the controls hooked up? Does it operate properly?

- If there are filters in hoods, are they installed and reasonable clean?
- If there are back draft dampers in the branch ducts or dampers in the hoods, make sure they are open.
- If there are safety controls in the system, check them.
- If there is a makeup air unit or fan, make sure it is running and pumping out the right quantity of air.

3. **Start Up The Fan:**
   - Take amp and volt reading
   - Read fan rpm
   - Check fan static pressure
   - Take pitot traverse of main duct for total flow

4. **Proportionate Balance:** If there is no diversity, start with the hood farthest from the fan, determine percent of design, read second to last hood and adjust so that it is the same percent of design as the last hood. Read third to last hood and set to within 0 and 5 percent of second to last hood. Proceed thusly hood by hood to fan.

5. **No Diversity:** The system may be designed for 100 percent usage of all the hoods at one time, in which case, all the branch dampers should be open. The fan CFM equals the sum of all the hoods in this case.

   Diversity: On other cases only a portion of the hoods may be used at any one time. Sometimes in actual practice the system is turned on and only one hood is used. Check the total CFM of all the hoods against the fan CFM for variations between the two.

   If the fan CFM is less than the hoods, balance with two sets of conditions. Close off the appropriate number of hoods nearest the fan first, test the open hoods, then close the end hoods, open up the rest and test again.

6. **Reread Amps and Static Pressure** at fan. Take final fan traverse and compare with hood total. Note if there is excessive infiltration into duct.

Chapter 7
VARIABLE AIR VOLUME CONVERSIONS

There are many potential areas where energy can be saved in variable air volume systems over constant volume operation.

A. Fan Savings

1. Fan electrical energy decreases with fan volume according to the cube fan law.

2. On an average during VAV operation, fans run at 40 percent less than full load during occupancy operation.

3. Fans in the past for HVAC systems were selected based on calculated total worst possible load concepts, which may on an average, be 25 percent too high.

   These peak calculated loads are generally too high for various reasons. All the parts of the building generally don't reach their peak loads simultaneously and hence will never need a fan running at that capacity. The calculated loads may carry design fudge factors of 5 to 20 percent, and various energy conservation methods in recent years may have reduced the load needed on the fan.

   Constant volume systems run constantly at these excessively high air volume loads and waste a great deal of energy.

Whereas, a supply fan in a VAV system running at 38,000 CFM will be automatically reduced possibly 30 percent by a 26,000 CFM actual peak load requirement.

4. When fans must run during unoccupied hours during nights and weekends due to weather conditions, piping freezing, skin loads etc., they can automatically run at their minimum capacities with VAV systems. This could well be a turndown of 75 percent of full load.

B. Cooling and Heating Equipment Savings

1. If less air passes over cooling and heating coils, less energy is needed to generate the cooling or heating.

2. Cooling and heating equipment can cycle on and off, by completely turned off for periods of times or cycle through stages commensurate with VAV variations in loads and generate large electrical savings.

3. The efficiency of the cooling and heating equipment can be increased and smaller equipment may be used with VAV systems taking advantage of diversity.

C. Reheat and Dual Duct VAV Conversion Savings

1. Variable air volume systems eliminate wasteful reheating and recooling of air already cooled or heated.
   There can be a savings of about 30 percent in heating costs in the frost belt states.
   Cooling costs savings can be in the area of 5 percent in the frost belt states and about 10 percent in the sun belt states by eliminating constant volume terminal reheat and dual duct systems.
   In terminal reheat and dual duct systems the heating must be in operation in the summer and cooling in operation during parts of winter. Converting to VAV will allowing turning this equipment off during these periods.

D. Pumps

1. If the heating equipment can be turned off in the summer and cooling equipment in the winter, so can their respective pumps, affording an excellent savings.

2. Smaller pumps and lower pumping capacities can be effected in VAV systems.

3. Variable volume pumps can be used.

E. Percentage Horsepower Savings with Different Methods of Fan Volume Control at an Average of 60 Percent of Peak Flow

	Maximum Savings
Backward Inclined Fans with Discharge Dampers	13%
Airfoil or BI Fans with Inlet Vanes	36%
Forward Curve Fans with Discharge Dampers Located 3 Fan Diameters from Fan	48%
Adjustable Speed Drive	50%
Forward Curve Fan with Inlet Vanes	57%
Adjustable Frequency AC Motor Control	78%

F. Ballpark Savings on VAV Energy Consumption

Total Building Energy Consumption, Percent Savings
- Range .... 20% to 50%
- Average 35%

Total Building Energy Costs Savings, Per Sq. Ft.
- Range .... $.30 to $.70
- Average .... $.50

Fan Savings .... 15% to 70%
Cooling Savings 30%
Savings on Pump Energy .... 10% to 15%
Heating Savings 30%

G. Budget Costs for Conversion

- $400 to $1200 per box

- 35¢ to $1.00 per sq. ft. of building
- $200 to $600 per ton

BUDGET ESTIMATING OPERATING COSTS

Budget Estimating Motor Operating Costs

Problems

- Fans Oversized
- Pumps Oversized
- Excessive Resistance in Air or Water Systems

The **extra yearly costs for one additional BHP** drawn on a motor based on $.08/kWh is:

$700/yr..full time
$350/yr..half time

The **extra yearly costs for one additional AMP** on a 230V 3-phase motor: $250/yr..full time
$125/yr..half time

Budget Estimating Cost of Extra Resistance in Systems

Problems

- Dirty components
- Old high resistance filters
- Dampers closed, etc.
- Poor ductwork or piping design or installation

The extra yearly costs due to additional resistance in ductwork or piping systems as follows is:

Piping.............**5 extra feet of heat = 1 extra BHP**
Ductwork........ **extra 1/2 inch static pressure = 1 extra BHP**
$700/yr per HP......full time at 8¢/kWh
$350/yr per HP.......half time at 8¢/kWh

Budget Estimating Boiler Operating Costs

Problems

- Operating at 40 to 70 percent combustion efficiencies
- Running at 40 to 70 percent partial load
- Dirty tubes

The **extra costs per year running 10 percent under design efficiency** based on 4800-hour winter avg 35°F outside temperature, running half the time, is:

	Gas	Oil	Electric
Per Million BtuH Input	$1,200	$1,700	$4,800

Budget Estimating Chillers Operating Costs

Problems

- Operating at 40 to 70 percent of full loads
- Running at lower efficiencies than design
- Dirty condenser tubes

The **extra costs for differences in kW per ton** over the design rating, based on 1000 hrs of operation and 8 cents per kWh electrical costs is:

$8/yr.................. per 0.1 kW/ton

Sample of operating costs per ton per year for various kW/ton ratings is:

		Loss
@ .7 kW/ton	$56/yr/ton	$0/yr/ton
@ .8 kW/ton	$64/yr/ton	$12/yr/ton
@ 1.0 kW/ton	$80/yr/ton	$24/yr/ton
@ 1.2 kW/ton	$94/yr/ton	$48/yr/ton

Hence, a 500-ton chiller at 1.2 kW/ton can lose $24,000 per year over a .7 kW/ton unit.

Budget Estimating Air Cooled DX Condensers Operating Costs

The **electrical costs** for running an air cooling condenser unit with a SEER rating of 8 during the summer at $.08 per kWh is:

1000 CFM..............about $100 per ton

Budget Estimating Lighting Energy Costs

General Problems with Lighting

Ten to fifth percent too must lighting provided in various areas.

Lighting not always turned off during unoccupied times

Inefficient lighting

Yearly Costs and Savings

Yearly savings reducing from 4W to 3W per sq. ft. of building, at 8¢/kWh and 4400 hr/yr operation.

@ 4W/sq ft	$1.41 sq ft/yr	
@ 3W/sq ft	1.06 sq ft/yr	
	$.35 sq ft/yr	savings

Hence, a 90,000 sq. ft. building reduced from 4 kW to 3 kW per sq. ft. of building would save 90,000 sq. ft. x $.35 = $31,500.

Budget Estimating Outside Air Energy Costs

Problems

Drawing in more minimum outside air than needed through the outside air intake or through infiltration when heating or cooling equipment is running.

The **extra costs per year to heat 1000 CFM of excessive minimum outside air** based on 4800 hour winter and avg 35°F outside air temperature is:

Full-time in winter per 1000 CFM (4800 hr, .75 eff.)

Gas..$1200/Yr
Oil...$1700/Yr
Electric....................................$3600/Yr

Half-time in winter per 1000 CFM (2400 hr, .75 eff.)

Gas..$600/Yr
Oil...$850/Yr
Electric....................................$1800/Yr

Full-time, summer 1000 hr, 75°F avg temp, per 1000 CFM

Electric Cooling............................$1450

Btu Heating Consumption and Costs Per Year For Different Size Buildings and Energy Consumption Rates

Building Size	Yearly Energy Consumption, Mill Btu's			Gas Costs at $5.00 per Million Btu
	Btu Per Sq Ft Per Year			
Sq Ft	25,000	50,000	75,000	
1,000	25	50	75	$250
2,000	50	100	150	500
3,000	75	150	225	750
4,000	100	200	300	1,000
6,000	150	300	450	1,500
8,000	200	400	600	2,000
10,000	250	500	750	2,500
15,000	375	750	1,125	3,750
20,000	500	1,000	1,500	5,000
25,000	625	1,250	1,875	6,250
30,000	750	1,500	2,250	7,500
35,000	875	1,750	2,625	8,750
40,000	1,000	2,000	3,000	10,000
50,000	1,250	2,500	3,750	12,500
70,000	1,750	3,500	5,250	17,500
90,000	2,250	4,500	6,750	22,500
100,000	2,500	5,000	7,500	25,000
150,000	3,750	7,500	11,250	37,500
200,000	5,000	10,000	15,000	50,000
300,000	7,500	15,000	22,500	75,000
400,000	10,000	20,000	30,000	100,000
500,000	12,500	25,000	37,500	125,000
600,000	15,000	30,000	45,000	150,000
700,000	17,500	35,000	52,500	175,000
800,000	20,000	40,000	60,000	200,000
1,000,000	25,000	50,000	75,000	250,000
2,000,000	50,000	100,000	150,000	500,000
3,000,000	75,000	150,000	225,000	750,000

Actual Heat Loss takes into consideration all heating losses and gains during winter.

Yearly Heating Costs Per Sq Ft of Roofs, Walls, Etc. For Different R Factors and Fuel Costs

Avg Transmission Factor of Roof, Wall		Heating Cost per Sq Ft of Roofs, Wall, Etc.						
		Fuel Cost Per Million Btu						
R	U	$4	$6	$8	$10	$15	$20	$25
0.88	1,130	$1.012	$1.514	$2.023	$2.531	$3.797	$5.062	$6.328
1.00	1.000	0.896	1.340	1.790	2.240	3.360	4.480	5.600
1.80	0.550	0.493	0.737	0.985	1.232	1.848	2.464	3.080
2.00	0.500	0.448	0.670	0.895	1.120	1.680	2.240	2.800
3.00	0.330	0.296	0.442	0.591	0.739	1.109	1.478	1.848
4.00	0.250	0.224	0.335	0.448	0.560	0.840	1.120	1.400
5.00	0.200	0.179	0.268	0.358	0.448	0.672	0.896	1.120
6.00	0.160	0.143	0.214	0.286	0.358	0.538	0.717	0.896
7.00	0.140	0.125	0.188	0.251	0.314	0.470	0.627	0.784
8.00	0.125	0.112	0.168	0.224	0.280	0.420	0.560	0.700
9.00	0.111	0.099	0.149	0.199	0.249	0.373	0.497	0.622
10.00	0.100	0.090	0.134	0.179	0.224	0.336	0.448	0.560
12.00	0.083	0.074	0.111	0.149	0.186	0.279	0.372	0.465
14.00	0.071	0.064	0.095	0.127	0.159	0.239	0.318	0.398
16.00	0.062	0.056	0.083	0.111	0.139	0.208	0.278	0.347
18.00	0.055	0.049	0.074	0.098	0.123	0.185	0.246	0.308
20.00	0.050	0.045	0.067	0.090	0.112	0.168	0.224	0.280
22.00	0.045	0.040	0.060	0.081	0.101	0.151	0.202	0.252
24.00	0.041	0.037	0.055	0.073	0.092	0.138	0.184	0.230
26.00	0.038	0.034	0.051	0.068	0.085	0.128	0.170	0.213
28.00	0.035	0.031	0.047	0.063	0.078	0.118	0.157	0.196
30.00	0.033	0.030	0.044	0.059	0.074	0.111	0.148	0.185
35.00	0.028	0.025	0.038	0.050	0.063	0.094	0.125	0.157
40.00	0.025	0.022	0.034	0.045	0.056	0.084	0.112	0.140
45.00	0.022	0.020	0.029	0.039	0.049	0.074	0.099	0.123
50.00	0.020	0.018	0.027	0.036	0.045	0.067	0.090	0.112

BASIS

1. **Based on: 4,800 hours winter seasons in northern half of the U.S., inside temperature of 70°F, average outside temperature of 35°F.**
2. **Based on Btu input to heating equipment with any efficiency of 75 percent.**

HEATING PICKUPS NOT COMPENSATED FOR:

1. **Motor operating heat pickup not included.**
2. **Pickup from lighting heat not included. To include multiply by .75.**
3. **People heating pickup not included.**

OTHER BUILDING LOADS

1. **To include average building electrical costs for HVAC equipment, motors and lighting per sq ft per year multiply the above per sq ft cost by 1.6. This covers all motors and lighting.**

PROCEDURE FOR EVALUATING A VAV CONVERSION

1. **Calculate the new peak heating and cooling loads** based on energy consumption measures planned or previously implemented and other possible system or building changes from original design. Distinguish true simultaneous peak building load from calculated building peak load, which assumes that the peak load occurs everywhere in a building at the same time. Not the diversity, which is the difference between the true simultaneous and calculated loads.

2. **Determine lower CFM required** based on new heating and cooling loads and other changes. Note diversity in CFM.

3. **Before taking existing readings on fan:**
   - Make sure filters, coils, etc. are reasonably clean.
   - Check system for imbalance. If excessive, open all outlets 100 percent.
   - Make sure automatic dampers are operating properly.
   - Check for duct leakage. Correct if excessive.
   - Make sure that the system is open and that no fire and manual dampers are shut.

4. **Take readings of actual existing fan performance.**
   - Total CFM flow
   - Suction and discharge fan static pressure
   - Fan RPM
   - Filter, coil, damper pressure drops
   - Outside air flow and operation
   - Running amps if in question
   - Record motor name plate data
   - Check starter and overload size

5. **Determine operational hours of fan.**

6. If possible **put amp recorder of fan** for 24 hours on a weekday and weekend day.

7. **Check the intake and discharge water temperatures, pressures and flows** at coils. If water flows and temperatures are excessively higher or lower than design, determine effect on air flow.

8. If working with **DX coils**, check if condenser system has unloaders, hot gas bypass or multiple compressors to avoid freeze up at coils with lower air volumes.

9. **Calculate new fan performance and drive sizes** based on new peak load. See Chapter 11, pages 000-000.

10a. **Select type of VAV system and VAV terminal units** best suited to building and existing HVAC systems.
Types of VAV systems available are:
  Cooling only, cooling with reheat coils
  Separate interior and perimeter systems
  Fan-powered terminals with or without reheat coils
  Bypass at terminal only or at both terminals and fan
  Dual duct VAV
  Introduction VAV
  Riding fan curve
See section in this chapter for details.

10b. The types of VAV terminals will correspond to the types of systems listed above. Further considerations for terminal selection are dependent on types of controls. See sections on VAV terminals and controls in this chapter for details.

10c. Make sure the duct design and outlet air distribution is suitable at maximum and minimum flows. Change outlets as required.

11. **Select type of fan volume control.**
Fan volume control may be implemented with:
  Inlet or outlet dampers
  Variable speed motor controls
  Variable speed drives
  Variable pitch vaneaxial fans, etc.
See section in this chapter for details.

12. **Select type of control system** best suited for new VAV system operation and compatibility with existing controls. See information on VAV controls in another section in this chapter.

13. **Calculate existing annual electrical consumption** of fan motor in terms of kWh and costs.

$$kWh = \frac{1.73 \times Amps \times Volts \times Hours}{1000}$$

$$= \frac{1.73 \times 77 \times 460 \times 5260}{1000}$$

$$= 322{,}070 \text{ kWh}$$

Old Costs = kWh x Costs/kWh = 322,070 x $.07 = $22,545

14. **Calculate fan savings.** The following calculations use system S-1 from the sample audit on the Suburban Office Building in Chapter 4.
Savings by Reducing Maximum CFM Load (See Chapter 11, p. 000, item 13.) The maximum CFM actually needed in the interior office areas served by system S-1 is a great deal less than being supplied. It can be reduced from about 1.2 CFM per sq ft to .84 CFM per sq ft. This reduces S-1 from 38,000 CFM to 26,000 CFM.

The fan savings due to this reduction are:

Existing	322,070 kWh	$22,545
New	113,000 kWh	$7,911
Savings	209,070 kWh	$14,634

Fan Savings Due to VAV Operation During Working Hours. The average reduction of air delivery in a VAV system during occupancy times is 40 percent of peak load. This reduces the maximum CFM of 26,000 for S-1 down to an average of about 15,600 CFM which is a reduction of 10,400 CFM.
The reduction in S-1 BHP, due to reducing the maximum CFM, is from the existing 59 BHP to 19 BHP (see p. 000).
The converted VAV system will run at a peak of 19 BHP and at an average of 60 percent of this during the 2,800 hours of occupancy.

BHP Savings	= .6 x 19 BHP = 11.40 BHP
Converted to kWh	= .746 x 11.40 = 8.5 kWh
Savings	= kWh x Hours x Cost/kWh
	= 8.5 x 2,800 x .07
	= $1,667 per year

Savings by running VAV boxes at minimum during evenings and weekends during winter months. Units were shut off during non-occupancy during summer.

BHP Savings at Minimums = .25 x 19 BHP = 4.75 BHP
Converted to kWh = .746 x $1.4 = 8.5 kWh
Savings 4 winter months (115 hrs/wk x 17 wks = 1955 hrs)
kWh x hrs x Cost/kWh = 8.5 x 1955 x .07 = $1,163

CFM SAVINGS WITH VAV SYSTEM

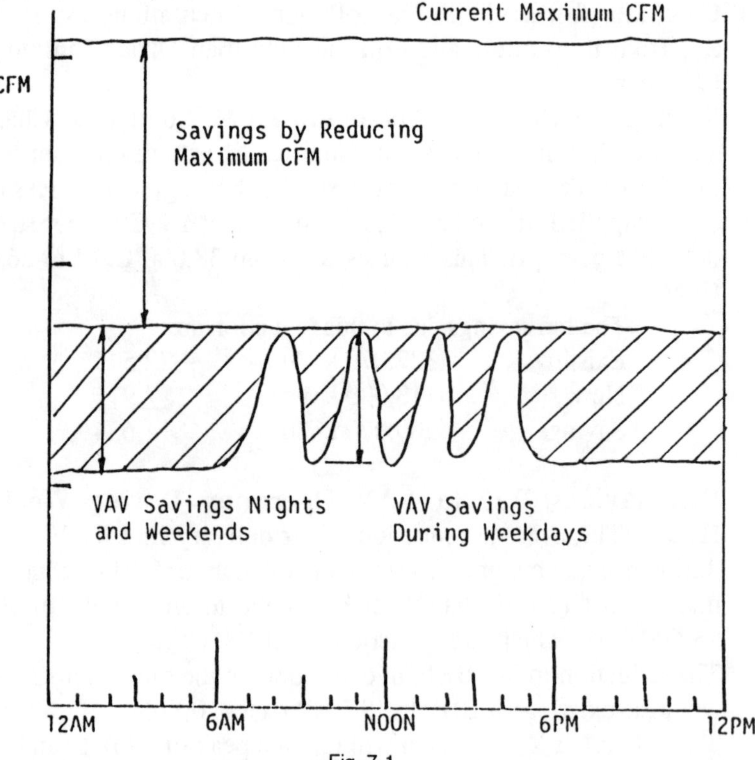

Fig. 7-1

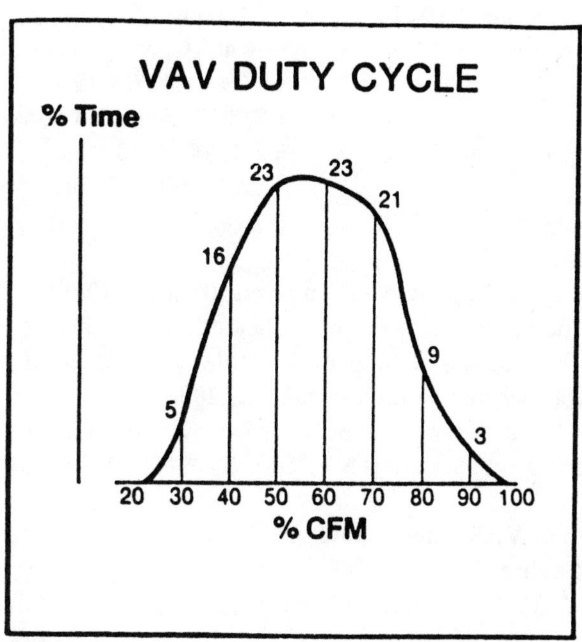

15. **Cooling Savings**

Cooling Savings by Reducing Maximum CFM Loads
Savings in cooling is mostly a sensible energy reduction. This more conservative calculation of savings will be taken:

CFM Reduction	= 38,000 – 26,000 = 12,000 CFM
BtuH Reduction	= 1.08 x CFM x TD
	= 1.08 x 12,000 x (74–55)
	= 246,240 BtuH
@ 60% avg load	= 246,240 x .60
	= 147,750 BtuH (12–3 Tons)
	= 389 mill Btu
Dollar Savings	= @ $20 per million Btu for elect.
	= $7,770

Cooling Savings by VAV Operation During Occupancy Times
Average Reduction of 40 percent of peak load during occupancy,

Average CFM Reduct.	= 26,000 CFM x .4
	= 10,400 CFM
	= 1.08 x 10,400 x 19
	= 213,400 Btu (17.8 Tons)
Occupancy Time	= 213,400 x 1400 hrs
	= @$20 per mill Btu electrical
	= $5,975

Cooling savings by Avoiding Simultaneous Cooling and Heating and the resultant recooling or heated air in reheat systems. S-1, S-2 represent 58% of the total cooling load which costs $66,000 per year which equals $38,280. There is a savings of 30 percent by eliminating unnecessary recooling which equals $38,280 x .3 = $11,484.

16. **Recap of VAV Energy Savings**

Fan Savings	
VAV Operation, Occupancy, S-1	$1,667
S-2	$1,667
Cooling Savings	
VAV Operation, Occupancy, S-1	$5,975
S-2	$5,975
Avoiding Recooling	$11,484
	$26,768

17. **Estimate Retrofit Costs**

Box conversion to VAV
Fan volume controls
Controls
Control center
Ductwork
Wiring
Piping
Total costs $75,500 (see Chapter 9)

18. **Calculate Payback**

$$\frac{\text{Costs}}{\text{Savings}} = \frac{\$75{,}500}{\$26{,}768} = 3.82 \text{ years}$$

19. If the VAV conversion is feasible, proceed with a detailed design, get quotations on equipment and installation and implement conversion.

20. Perform proper start-up, check out performance, test and balance and monitor.

ENERGY AUDITING PROCEDURES

The purpose of an energy audit is to determine the energy consumption and costs of the overall building and of its specific components, the structure, systems and equipment. It aims to generate energy improvement options, to project energy savings, to estimate the costs of energy improvements, calculate paybacks, and on this basis evaluate the various options.

A good audit is diagnostic in nature, develops a valid prognosis of the causes of energy wastes, and leads to scientifically established remedies.

There are two basic phases or types of audits, **short walk-through audits** and **in-depth detailed audits**, either of the entire building or of only selected parts of building.

Phase One: Walk-Through Energy Audit Procedure

1. Make an initial walk-through inspection to become familiar with the building, systems, equipment, maintenance, operation status, etc.
2. Study at the plans and specifications and become familiar with the building, systems, capacities, equipment, etc.
3. Talk briefly with the building operating personnel, owner, occupants etc. about HVAC systems, comfort, problems, etc.
4. Examine the overall building energy consumption history from the owner if available. If not, get complete energy consumption history on gas, oil, electrical, etc., from utility companies and fuel suppliers.

   Compare the Btu consumption per sq. ft. per year with other similar buildings and determine degree of variance.

5. List maintenance, cleaning, adjustment, repairs and balancing needed to this point. Determine what maintenance and repairs must be done before the detailed audit can be performed.
6. Take some spot test readings if needed.
7. If a more extensive audit is needed, determine what test readings, inspections, analysis, calculations, etc. are required and estimate the time and costs involved.
8. Fill out Building and Systems Description report.
9. Write out a list of existing energy problems.
10. List obvious and potential energy savings improvements. Develop the most promising energy improvements further.
11. If the walk-through audit is sufficient, calculate energy savings for the various energy improvements, estimate retrofit costs and calculate paybacks.
12. Select with owner which energy improvements to proceed with and assign priorities. Properly engineer retrofit work and proceed.

Average Annual Energy Performance In Btu's Per Square Feet

		Heating and Cooling Degree Day Region						
Building Type	National	1	2	3	4	5	6	7
Office	84,000	85,000	76,000	65,000	61,000	51,000	50,000	64,000
Elementary	85,000	114,000	70,000	68,000	70,000	53,000	48,000	57,000
Secondary	52,000	77,000	65,000	55,000	51,000	37,000	41,000	34,000
College/Univ.	65,000	67,000	70,000	46,000	59,000			83,000
Hospital	190,000		209,000	171,000	227,000	207,000		197,000
Clinic	69,000	84,000	72,000	71,000	65,000	61,000	59,000	59,000
Assembly	61,000	58,000	76,000	68,000	51,000	44,000	68,000	57,000
Restaurant	159,000	162,000	178,000	186,000	144,000	123,000	137,000	137,000
Mercantile	84,000	99,000	98,000	86,000	81,000	67,000	83,000	80,000
Warehouse	65,000	75,000	82,000	65,000	50,000	38,000	37,000	39,000
Residential Non-Housekeeping	95,000	99,000	84,000	94,000	125,000	90,000	93,000	106,000
High Rise Apt.	49,000	53,000	53,000	52,000	53,000	84,000	20,000	

February 1978, HUD-PDR-290

(See map next page)

Average Annual Energy Regions

REGION 1 REGION 2 REGION 3 REGION 4

REGION 5 REGION 6 REGION 7

PHASE TWO:
IN-DEPTH ENERGY AUDIT PROCEDURE

Field Surveys

1. Make thorough inspection of building systems and equipment and become thoroughly familiar with them. Check out operations, performance, maintenance, malfunctions, comfort, problems, etc.

2. Check name plate data on equipment.

3. Conduct in-depth interviews with building personnel. Review maintenance, scheduling, performance, comfort and problems of building, equipment and systems.

4. Become familiar with actual hours of operation of systems and equipment, and the hours of occupancy of the personnel.

Energy History

5. Study and analyze a 3-year history of the building's electrical and fuel energy consumption. Compare with building consumption indexes of similar buildings.

Field Tests

6. Take test readings of actual flows, temperatures, pressures, rpm's, amps, volts, etc., at HVAC equipment.
   Check pressure drops across filters, coils, strainers, etc. Check outside air flows at minimum and maximum.
   Monitor readings over a period of time with a recording instrument where required.
   Check lighting levels.

Seasonal and Peak Energy Calculations

7. Determine the actual existing seasonal and peak energy consumption and efficiencies of specific systems and equipment, etc. based on tests and other data.

8. Calculate the peak and seasonal heating, cooling and CFM loads actually required to meet current conditions for the overall building and various areas of the building. Compare with the design and existing capacities.

Evaluation of Energy Improvements

9. List all problems with building, systems and equipment.

10. Generate energy improvements and develop those with most potential. Write out list of improvements.

11. Calculate the potential energy savings in terms of Btu's and kWh's, and in costs.

12. Estimate costs of retrofit work.

13. Calculate paybacks and return on investments.

IN-DEPTH ENERGY AUDIT PROCEDURE

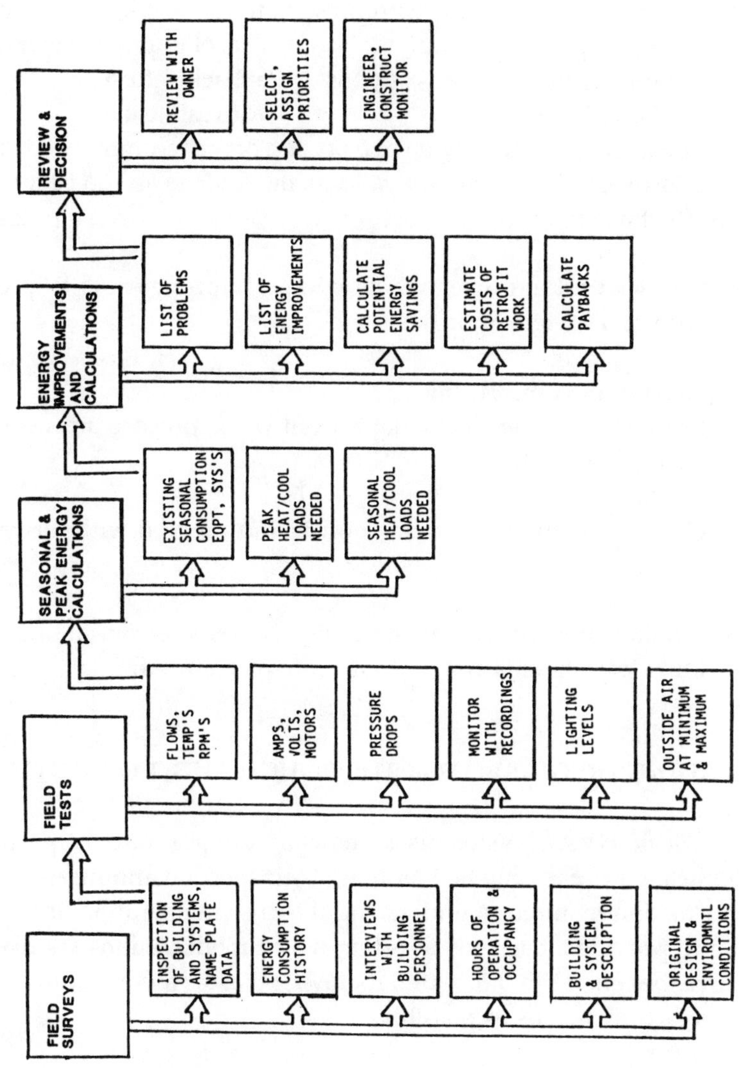

Review and Decisions

14. Review with owner or his representatives:

Problems	Costs of improvements
Energy improvement options	Payback, return on investment
Potential savings	

    Consider a change only on one portion of the recommended energy improvements to test and validate the savings and to observe the effects.

15. Select with the owner which energy improvements to proceed with and assign priorities.

Engineering and Construction

16. Properly engineer the owner retrofit work, prepare drawings and write specifications.

17. Obtain quotations, award contracts and proceed with the retrofit work.

18. Monitor units of energy and costs savings after put in operation. Make adjustments and modifications as required.

READINGS REQUIRED FOR THOROUGH AUDIT OF SYSTEM

Each HVAC system is somewhat unique and its particular characteristics can only be identified by inspection and measurement. Information required to understand the present operation of a system and to provide a basis for deciding which modifications are likely to prove beneficial is tabulated below. Measurement should be as nearly simultaneous as possible.

Electrical Readings (Amps. Volts, Power Factors, kW)

- Fans
- Pumps
- Compressors
- Condensors
- Chillers

	Lights Owners operating equipment
Air Flow Rates	Total supply air from fan Total return air to fan Total outdoor air Trunk ducts Terminal units Air cooled condenser
Air Pressure Readings	Suction and discharge of fans Drops across coils, filters etc.
Water Flow Rates	From pumps Through boilers Through chillers Cooling towers Heat exchangers Coils and terminal units
Water Pressure Readings	Suction and discharge of pumps Drops across strainers, coils etc. Drops across boilers, chillers, condensers
Temperatures, Air	Outdoor air DB & WB Return air DB & WB Mixed air entering coils, DB & WB Supply air leaving coils, DB & WB Hot deck Cold deck Air at terminals Conditioned areas DB & WB
Temperatures, Water	Boiler supply and return Chiller supply and return Condenser supply and return Heat exchanger supply and return Coil supply and return

Refrigerant Temperatures	Hot gas line Suction line
Overall Building Energy Readings	At gas meter with all heating on At electric meter with only lights on At electric meter with HVAC units on At electric meter with refrigeration on

Energy conservation must be approached in a systematic manner rather than considering individual items out of context. Systems do not operate in isolation but depend on and react with other systems. It is important to recognize this interaction of systems as modification to one will cause a reaction in another which may be either beneficial or counter-productive.

SAMPLE AUDIT AND FORMS

SUBURBAN OFFICE BUILDING

This 90,000-sq-ft suburban office building was built around 1967 before the energy crisis occurred and it incorporates in its design and operation the great energy waste of that era. Overall, the HVAC, electrical and plumbing systems in the building consumed 276,000 Btu per sq. ft. per year for a total consumption of 24.9 bill Btu for 1984. Energy costs were $3.80 per sq. ft. and totaled $339,800 for the year.

The targeted energy reduction is 50 percent, reducing the Btu per sq. ft. from 276,000 down to 138,000 Btu per sq. ft. with savings of about $1.90 per sq. ft. or $171,000 per year.

Problems with Existing HVAC System and Building

1. The building has two energy wasting HVAC systems which simultaneously heat and cool, an interior high pressure terminal reheat system and a high pressure perimeter induction system with about one third primary air taken from the outside.

2. The 310-ton chiller, 6,000,000 BtuH boiler and the hot and chilled water pumps must run the year around because of the terminal reheat system, wasting a great deal of energy.

3. The computer room operates 24 hours a day, weekdays, with a constant cooling load demand forcing the chiller to run the year around and sporadically at nights.

4. The chiller and boiler are oversized; they cycle and run at inefficient levels.

5. Excessive minimum outside air is brought into the building. The settings of the dampers are off and they leak.

6. The lighting levels are excessive.

7. Thermostats are set too high or low in winter and too low in summer.

8. The oil-fired boiler is inefficient, scaley and has a poor combustion efficiency of 60 percent.

9. Maintenance is poor. Filters, coils and strainers are generally dirty. Many control valves, automatic dampers and thermostats are out of calibration, malfunctioning or mis-set.

10. Systems are out of balance.

11. Fans are oversized for load, pumping out more air than required and running 24 hours per day year around.

12. Pumps are oversized for load, pumping out more GPM's than required, running on demand 24 hours a day year around.

13. Starting and stopping times of HVAC equipment not optimized.

14. Paying higher demand rates than need be.

15. Power factors on underload motors not controlled.

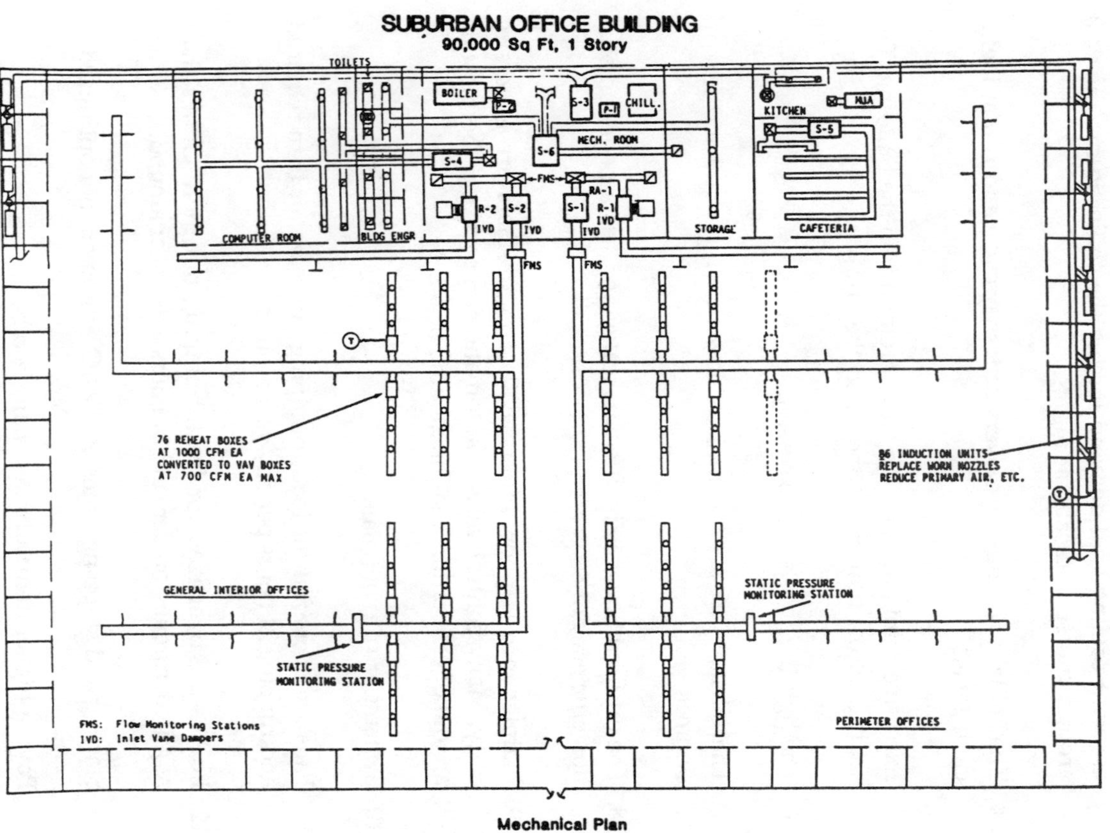

Mechanical Plan

BUILDING AND SYSTEM DESCRIPTION

Name Suburban Office Building

Location

Latitude 41°N Elevation 658 Ft When Built 1967

A. CATEGORY OF STRUCTURE

Office Building

B. BUILDING DESCRIPTION

Area, Sq Ft: 90,000 Number of Floors: 1

Volume, Cu Ft: 1,260,000

Number of Occupants: 400 Sq Ft/Person: 225

Types of Areas: Offices, computer room, kitchen, dining room, employee lounges, storage, mechanical room

C. CONSTRUCTION DETAILS

Glass: Single pane, clear U=1.13, no shading, no drapes or blinds, sealed, aluminum frame

Exterior Walls: 8" brick and block, lathe and plaster, R-6, U factor .167

Roof and Ceilings: Built up tar and gravel on 2" rigid insulation (R-6 & metal deck, suspended accoustical ceiling. U factor .11

Floors: Concrete slab, 2 BTUH per sq ft

Total Exposed Wall Area Sq Ft: 10,430

Total Glass Area Sq Ft: 6,650 Percent 39

D. HOURS OF OCCUPANCY AND OPERATION

Working Hours: 8am to 6pm weekdays, 9am to noon sat.

Lighting Hours: 8am to 9pm weekdays, 3 hrs sat.

HVAC Hours: Cooling and heating year around

Janitorial Cleanup Times: 6pm to 9pm weekdays

Computer Room: 24 hrs per day, 365 days per year

Other:

E. HEATING AND COOLING SYSTEMS DESCRIPTION

Interior office spaces, high pressure terminal reheat system
High pressure perimeter induction system, 100% primary air
Chilled water cooling: 1 centrifugal chiller
Hot water heating, oil fired
Kitchen, Dining: Single zone low pressure
Computer Room: Single zone low pressure

F. ANNUAL ENERGY CONSUMPTION

Total Heating, Cooling, Electrical, Lighting Per Yr:

Total BTU: 24.9 Bill. BTU Per Sq Ft: 276,600

Total Energy Costs: $339,862 Costs Per Sq Ft: $3.78

Electrical, Total KWH: 3,829,411 KWH/Sq Ft: 42.55

Total Elec. Costs: $268,062 Costs/Sq Ft: $2.98

Heating Fuels, BTU Per Yr: 11.8 Bill. Per Sq Ft: 131,110

Total Fuel Costs: $71,800 Costs/Sq Ft: $.81

G. ORIGINAL ENVIROMENTAL DESIGN CONDITIONS

Heating

Peak Heat Loss BTUH: 4.8 Mill. Output Degree Days: 6,000

Design Temperatures: -10F, 74F

Avg Winter Temp.: 35 Avg Winter Hours: 4,800

Cooling

Peak Heat Gain BTUH: 310 tons Degree Days: 682

Design Temperatures: 94F DBT, 75F WBT Outdoors
74F DBT 50% RH Indoors

Avg Summer Temp.: 75 Avg Summer Cool Hours: 900

Air and Hydronic Flows

Supply CFM: 121,000 CFM/Sq Ft: 1.34

Exhaust Air CFM: 7,300 Exh Air/Sq Ft: .08

Min Outside Air CFM: 30,000 OA Per Person, Sq Ft: 75 or .3/sq ft

Make Up Air CFM: 5,400

HVAC GPM: 1,954 Domestic GPM:

H. LIGHTING

Levels in Foot Candles: 100-200

Levels in Watts/Sq Ft: 4.0 avg.

Type: Fluorescent

I. ELECTRICAL SERVICE

Type: Underground Metering: Primary

Voltage: 277/480V, 3 phase, 4 wire, wye

J. CONNECTED ELECTRICAL LOADS (KW's)

Lighting: 360 KW Office Equipment: 37+10 comp KW

Heating and Cooling Equipment: 286 KW

Air Handling and Exhausts: 160 KW

Cooking: 50 KW Machinery:

Total: 903 KW

HVAC EQUIPMENT SCHEDULE

Job: Corp. Office Bldg. Location: Date: 11-8-84

SYSTEM	SERVES	EQPT LOCAT	TYPE EQPT	☒ FAN ☒ CFM ☐ GPM	☐ PUMP RPM	☒SP ☐FT HD	OUTLET VELOC	MOTOR HP	AMPS	VOLTS	COOLING COIL MBH	TONS	GPM	HEATING COIL MBH	GPM
S-1	Interior Office East		AHU	38.000	1585	8"		75	96	460	1140	95	228	800	80
S-2	Interior Office West		AHU	38.000	1585	8"		75	96	460	1140	95	228	800	80
S-3	Perimeter Offices		AHU	7.000	2832	7"		10	28	230	600	50	120	760	76
S-4	Computer Rm.		AHU	10.400		2"		10	28	230	312	26	62	125	13
S-5	Cafeteria & Kitchen		AHU	9.800		1½"		5	15	230	300	25	48	70	70
S-6	Toils. Stor. Mech. Rm.		MZ	7.200		1½"		5	15	230	216	18	43	89	89
	TOTALS			121.200				180			3708	303	729	2644	408
MUA-1	Kitchen		MUA	5,400	760	1"		3	9	230	---	---	---	467	---
	Motors:		3 P.H..,	1 Cycle.	1750 RPM										

FAN EQUIPMENT SCHEDULE

Job: Corp. Office Bldg Location: Date: 11-8-84

SYSTEM	SERVES	EQPT LOCAT	TYPE EQPT FANS	☒ FAN ☒ CFM ☐ GPM	☐ PUMP RPM	☒SP ☐FT HD	OUTLET VELOC	MOTOR HP	AMPS	VOLTS	COOLING COIL MBH	TONS	GPM	HEATING COIL MBH	GPM
R-1	Interior Office East		Centrif	32.000	509	2"		15	42	230					
R-2	Interior Office West		Centrif	32.000	509	2"		15	42	230					
TE	Toilets		Pre	675	700	3/8"		1/4	3	115					
KE	Kitchen Hood		Pre	5.400	760	1"		3	9	230					
E-1	Conference Rm		Wall	600	960	1/4"		1/4	3	115					
E-2	Conference Rm		Wall	600	960	1/4"		1/4	3	115					
	TOTALS			71.275				34							

PUMP EQUIPMENT SCHEDULE

Job: Corp. Office Bldg Location: Date: 11-8-84

SYSTEM	SERVES	EQPT LOCAT	TYPE EQPT	☐ FAN ☐ CFM ☒ GPM	☒ PUMP RPM	☐SP ☒FT HD	OUTLET VELOC	MOTOR HP	AMPS	VOLTS	COOLING COIL MBH	TONS	GPM	HEATING COIL MBH	GPM
P-1	Chilled Water		Pump	744	1750	93		30	40	460					
P-2	Hot Water		Pump	466	1750	80		20	27	460					
P-3	Cooling Tower		Pump	744	1750	93		30	40	460					
CH-1	Chiller							300	339	460		310	744		

(224w)

PROPOSED ENERGY-SAVING RENOVATIONS

1. Perform a thorough detailed audit with complete test readings, engineering calculations and pricing, after which decisions are made on what retrofit work will be estimated and performed.

2. The reheat system will be converted to variable air volume (VAV). Some reheat coils will be eliminated. An energy management system will be installed to automatically control fan volume etc.

3. The induction system fan will be turned off in winter and run at a reduced rate only during occupancy during summer.

4. Lighting levels will be reduced from 4 watts per sq. ft. to 3.

5. Winter space thermostat settings to be reset from 74°F to 70°F and summer from 74°F to 78°F. All stats to be checked and recalibrated if needed.

6. Fan CFM's will be reduced by reducing rpm's.

7. Pump GPM's will be reduced by replacing pump.

8. Minimum outside ventilation air will be reduced from 30,000 CFM to 13,000 CFM.

9. All filters, coils, strainers, condenser tubing and boiler tubing to be checked and cleaned as required.

10. Change computer room unit from a CHW to DX system and reduce the tonnage from 26 to 18 tons. Also add an economizer section onto the unit and include staging in the condenser.

11. The boiler will be cleaned, checked and tuned up.

12. Install computer controlled optimizer on chiller.

13. Install automatic combustion controls and temperature reset controls on boiler.

14. An energy management system will be installed to control on and off times of lighting, fans, HVAC units, pumps, chillers and to minimize demand costs.

15. The systems will be rebalanced, air and water, for maximum efficiency of operation and optimum minimum energy consumption.

16. The systems will then be monitored to keep track of consumption and adjustments and modifications made as required.

Electrical Consumption History

Building Suburban Office Building Year 1984

Size Sq Ft 90,000

Month	No. of Days	kWh Used	Cost per kWh	Electrical Costs Demand Peak	Charge	Power Factor Adj.	Fuel Adj.	Total Cost
Jan.		262,651	0.0670		$769		0.200	$17,598
Feb		285,739	0.0600		659		0.200	$17,144
Mar.		219,792	0.0722		769		0.417	$15,875
Apr.		230,782	0.0670		659		0.417	$15,462
May		311,066	0.0760		1,429		0.200	$23,636
June		362,657	0.0780		1,429		0.200	$28,287
July		403,318	0.0720		1,319		0.200	$29,039
Aug.		429,693	0.0720		1,429		0.200	$30,938
Sept.		422,001	0.0750		1,539		0.200	$31,650
Oct.		350,568	0.0690		1,319		0.200	$24,189
Nov.		275,839	0.0620		769		0.200	$17,102
Dec.		270,161	0.0640		659		0.200	$17,290
Total		3,824,207	0.0701	0	12,748			$268,210
Avg./Mo		318,684	0.0701	0	1,062			$22,351

1. Average kWh per sq. ft. of building per year $42.49
   Average electrical cost per sq. ft. of building per year $2.98
2. BtuH equivalent for year, (kWh = 3413 BtuH) 13.05 billion
   BtuH average per month 1.09 billion
3. Average cost per million BtuH $20.55
   Average Btu per sq. ft. of building per year for electrical 13,052
4. Average kWh per hour 437

ELECTRICAL LOADS

☒ EXISTING ☐ NEW

BUILDING SUBURBAN OFFICE BUILDING Date

AIR HANDLING EQUIPMENT ☒ Seasonal ☐ Summer ☐ Winter ☒ Year Around

ITEM	HP OR KWH	RATED VOLTS	RATED AMPS	RATED PH	HOURS OF OPERATION PER YEAR	ACTUAL AVERAGE VOLT LOAD	ACTUAL AVERAGE AMP LOAD	KWH PER YEAR	COSTS PER YEAR
S-1, AHU, INT. EAST	75	460	96	3	5,260	460	77	322,315	$22,562
S-2, AHU, INT. WEST	75	460	96	3	5,260	460	77	322,315	$22,562
S-3, AHU, PERIMETER	10	460	28	3	5,260	460	25	104,648	$7,325
S-4, AHU, COMPUTER RM	10	460	28	3	8,760	460	25	174,280	$12,200
S-5, AHU, CAFETERIA	5	460	5	3	2,500	460	4	7,958	$557
S-6,MZU,TOILETS,STORAGE	5	460	5	3	8,760	460	4.2	29,279	$2,050
MUA-1, KITCHEN	3	230	9	3	2,500	230	7	6,963	$487
R-1, CENTRIFUGAL FAN	15	230	42	3	5,260	230	35	73,253	$5,128
R-2, CENTRIFUGAL FAN	15	230	42	3	5,260	230	34	71,160	$4,981
PRE, TOILET EXHAUST	0.25	115	3	1	8,760	115	2	3,486	$244
PRE, KITCHEN EXHAUST	3	230	9	3	2,600	230	8	8,276	$579
E-1, WALL FAN, CONF.	0.25	115	3	1	2,600	115	2.5	1,293	$91
E-2, WALL FAN, CONF.	0.25	115	3	1	2,600	115	2.5	1,293	$91
(S-1,S-2,S-3,R-1,R-2 off nights and weekends,April thru Nov.)									
TOTAL	217		369				303.2	1,126,520	$78,856

$$\text{KWH/YEAR} = \frac{1.73 \times I \times E \times \text{HOURS}}{1000}$$

COSTS PER KWH = $0.070

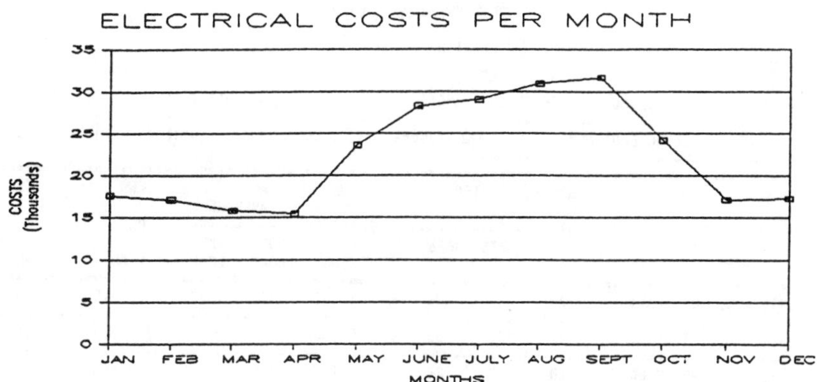
ELECTRICAL COSTS PER MONTH
COSTS (Thousands)
35
30
25
20
15
10
5
0
JAN FEB MAR APR MAY JUNE JULY AUG SEPT OCT NOV DEC
MONTHS

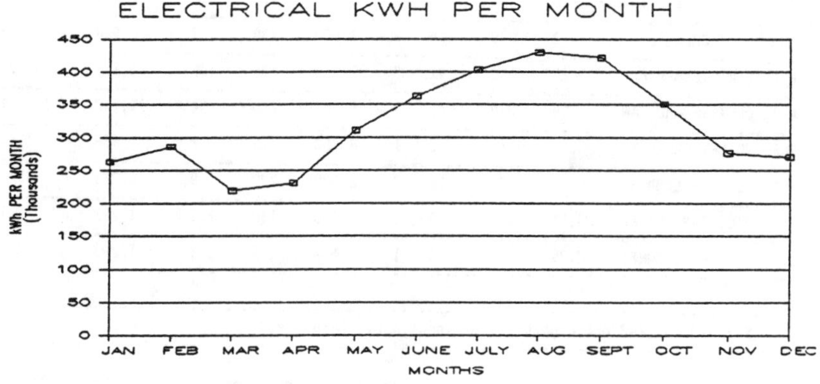
ELECTRICAL KWH PER MONTH
kWh PER MONTH (Thousands)
450
400
350
300
250
200
150
100
50
0
JAN FEB MAR APR MAY JUNE JULY AUG SEPT OCT NOV DEC
MONTHS

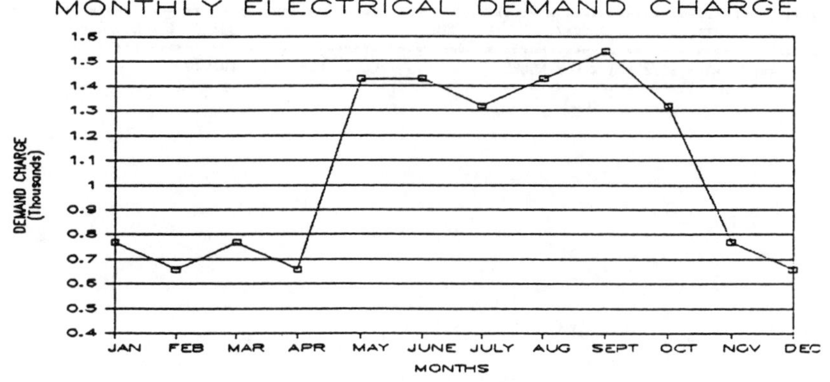
MONTHLY ELECTRICAL DEMAND CHARGE
DEMAND CHARGE (Thousands)
1.6
1.5
1.4
1.3
1.2
1.1
1
0.9
0.8
0.7
0.6
0.5
0.4
JAN FEB MAR APR MAY JUNE JULY AUG SEPT OCT NOV DEC
MONTHS

ELECTRICAL LOADS

☒ EXISTING ☐ NEW

BUILDING SUBURBAN OFFICE BUILDING Date 09-27-1985

AIR HANDLING EQUIPMENT ☒ Seasonal ☐ Summer ☐ Winter ☒ Year Around

ITEM	HP OR KWH	RATED VOLTS	RATED AMPS	RATED PH	HOURS OF OPERATION PER YEAR	ACTUAL AVERAGE VOLT LOAD	ACTUAL AVERAGE AMP LOAD	KWH PER YEAR	COSTS PER YEAR
S-1, AHU, INT. EAST	75	460	96	3	5,260	460	77	322,315	$22,562
S-2, AHU, INT. WEST	75	460	96	3	5,260	460	77	322,315	$22,562
S-3, AHU, PERIMETER	10	460	28	3	5,260	460	25	104,648	$7,325
S-4, AHU, COMPUTER RM	10	460	28	3	8,760	460	25	174,280	$12,200
S-5, AHU, CAFETERIA	5	460	5	3	2,500	460	4	7,958	$557
S-6,MZU,TOILETS,STORAGE	5	460	5	3	8,760	460	4.2	29,279	$2,050
MUA-1, KITCHEN	3	230	9	3	2,500	230	7	6,963	$487
R-1, CENTRIFUGAL FAN	15	230	42	3	5,260	230	35	73,253	$5,128
R-2, CENTRIFUGAL FAN	15	230	42	3	5,260	230	34	71,160	$4,981
PRE, TOILET EXHAUST	0.25	115	3	1	8,760	115	2	3,486	$244
PRE, KITCHEN EXHAUST	3	230	9	3	2,600	230	8	8,276	$579
E-1, WALL FAN, CONF.	0.25	115	3	1	2,600	115	2.5	1,293	$91
E-2, WALL FAN, CONF.	0.25	115	3	1	2,600	115	2.5	1,293	$91
(S-1,S-2,S-3,R-1,R-2 off nights and weekends,April thru Nov.)									
TOTAL	217		369				303.2	1,126,520	$78,856

$$KWH/YEAR = \frac{1.73 \times I \times E \times HOURS}{1000}$$

COSTS PER KWH = $0.070

ELECTRICAL LOADS

☒ EXISTING ☐ NEW

BUILDING SUBURBAN OFFICE BUILDING Date 09-27-1985

HEAT, COOLING EQUIPMENT AND PUMPS ☒ Seasonal ☐ Summer ☐ Winter ☒ Year Around

ITEM	HP OR KWH	RATED VOLTS	RATED AMPS	RATED PH	HOURS OF OPERATION PER YEAR	ACTUAL AVERAGE VOLT LOAD	ACTUAL AVERAGE AMP LOAD	KWH PER YEAR	COSTS PER YEAR
P-1, PUMP, CHW	30	460	40	3	3,180	460	32	80,981	$5,669
P-2, PUMP, HW	20	460	27	3	3,180	460	22	55,674	$3,897
P-3, PUMP, CONDENSER	30	460	40	3	3,180	460	32	80,981	$5,669
CHILLER	310	460	339	3	3,180	460	290	733,887	$51,372
COOLING TOWERS	9	230	28.8	3	3,180	230	23	29,102	$2,037
TOTAL	399		474.8				399	980,625	$68,644

$$\text{KWH/YEAR} = \frac{1.73 \times I \times E \times \text{HOURS}}{1000}$$

COSTS PER KWH = $0.070

ELECTRICAL LOADS RECAP

☒ EXISTING ☐ NEW

BUILDING SUBURBAN OFFICE BUILDING Date 11-18-1985

☒ Year Around ☐ Summer ☐ Winter ☐ Peak

ITEM	CONNECTED LOAD KWH	CONNECTED LOAD HP	PH	HOURS OF OPERATION PER YEAR	ACTUAL LOAD KWH	ACTUAL LOAD BHP	ACTUAL KWH PER YEAR	COSTS PER YEAR	PERCENT OF TOTAL
LIGHTING, WINTER	360			2,200	360		792,000	$55,440	0.21
SUMMER	360			1,800	360		648,000	$45,360	0.17
AIR HANDLING EQPT.	160	216					1,126,520	$78,856	0.29
CHILLERS,PUMPS,COOL.TWRS	286	384					980,625	$68,644	0.26
COMPUTERS	10			8,760			87,600	$6,132	0.02
OFFICE EQUIPMENT	37	25		2,600			65,000	$4,550	0.02
KITCHEN EQUIPMENT	50			2,500			125,000	$8,750	0.03
3,829,411 KWH / 90,000 sq ft			=	42.55 KW / sq ft / yr					
$268,062 / 90,000 sq ft			=	$2.98 / sq ft / yr					
3,829,411 KWH x 3416 BTU / KW			=	13.1 Billion equivalent BTU					
TOTAL	903	625					3,824,745	$267,732	1.00

1 HP = 746 WATTS

COSTS PER KWH = $0.070

PEAK HEATING, COOLING AND CFM PER AREA

☒ EXISTING ☐ NEW

BUILDING SUBURBAN OFFICE BUILDING DATE

TOTAL SQ FT 90,000 AVERAGE CFM PER TON 400

AREA	SQ FT OF AREA	COOLING LOAD		HEATING LOAD		CFM SUPPLY		DIRECT EXT. CFM
		SQ FT PER TONS	TONS	BTU PER SQ FT	MBH	CFM PER SQ FT	CFM	
INTERIOR OFFICES, EAST	31,275	330	95	25	782	1.20	37,909	
INTERIOR OFFICES, WEST	31,275	330	95	25	782	1.20	37,909	
PERIMETER OFFICES, EAST	3,489	250	14	60	209	1.60	5,582	
PERIMETER OFFICES, SOUTH	4,950	225	22	60	297	1.60	8,800	
PERIMETER OFFICES, EAST	3,489	250	14	60	209	1.60	5,582	
CONFERENCE ROOM, EAST	450	250	2	60	27	1.60	720	1,200
CONFERENCE ROOM, WEST	450	250	2	60	27	1.60	720	1,200
VESTIBULE	225	100	2	35	8	4.00	900	
COMPUTER ROOM	3,600	140	26	35	126	2.90	10,286	
MENS TOILETS	224	400	1	53	12	1.00	224	225
WOMENS TOILETS	224	400	1	53	12	1.00	224	225
WOMENS LOUNGE	224	400	1	53	12	1.00	224	225
HALLWAY	300	400	1	35	11	1.00	300	
BUILDING ENGINEER	225	400	1	35	8	1.00	224	
STORAGE ROOM	1,800	650	3	35	63	0.61	1,108	
MECHANICAL ROOM	4,800	400	12	35	168	1.00	4,800	
KITCHEN	1,000	143	7	60	60	2.80	2,797	5,400
CAFETERIA	2,000	150	13	35	70	2.70	5,333	
TOTAL	90,000	290	310	31	2,883	1.37	123,642	8,475

HEAT LOSS CALCULATION

☐ PEAK PER HR ☒ SEASONAL ☒ EXISTING ☐ NEW

Building SUBURBAN OFFICE BUILDING Date 11-18-1985

Location Latitude 41

Type Building OFFICE AND LABS Stories 1 When Built 1967

Sq Ft Area 90,000 Cubic Ft of Space 1,260,000

X Calculation for Whole Building For Partial Area

Budget Load: Sq Ft 90,000 x BTUYR/Sq Ft 78,500 = 7,065 Mill BTUYR

Outside Design: DB -10 WB Avg OA Temp. 35 Winter Hours 4,800

Inside Design: DB(day) 74 RH DB(night)

ITEM	DIMENSIONS	SQ FT	U	TEMP DIFF	BTU: Per Hour Seasonal	SEASONAL HOURS
ROOF OR CEILING	360 X 250	90,000	0.110	39.00	1,853,280,000	4,800
FLOOR	360 X 250	90,000	0.100	16.00	691,200,000	4,800
GLASS	950 X 7	6,650	1.100	39.00	1,369,368,000	4,800
DOORS						
WALLS	1220 X 14	10,430	0.330	39.00	644,323,680	4,800
COLD INSIDE WALLS						
VENTILATION	1.43 AC/HR CFM=	30,000	1.080	39.00	6,065,280,000	4,800
DUCT LOSSES						
GROSS TOTAL HEAT LOSS					10,623,451,680	
HEAT GAINS; LITES	4W/SQ FT	90,000	4.000	3.416	3,074,400,000	2,500
PEOPLE	No. People =	400	250		200,000,000	2,000
OFF. EQUIPMENT	HP =	25	2,550		127,500,000	2,000
COMPUTERS	KW =	5		3416.00	149,620,800	8,760
TOTAL INTERNAL GAINS					3,551,520,800	
NET TOTAL BUILDING HEAT LOSS					7,071,930,880	78,577*
INPUT TO HEATING EQUIPMENT, efficiency	=		.60		11,800,000,000	131,110*

BTUH = Sq Ft x U x Temp. Diff. BTU/YEAR = Sq Ft x U x Avg Temp. Diff. x Winter Hours

* BTU/SQ FT/YEAR

Remarks ______________________________

COOLING LOAD CALCULATION

☒ PEAK PER HR ☐ SEASONAL ☒ EXISTING ☐ NEW

Building SUBURBAN OFFICE BUILDING Date

Location Latitude 41 Peak Load, HR, MO:

Type Building OFFICE AND LABS Stories 1 When Built 1967

Sq Ft Area 90,000 Cubic Ft of Space 1,260,000

X Calculation for Whole Building For Partial Area

Building Load: Sq Ft 90,000 x Sq Ft/Ton 270 = 333 Tons

Outside Design: DB 95 WB 75 Avg OA Temp. 35 Summer Hours 900

Inside Design: DB(day) 74 RH DB(night) 74

ITEM (Orient.)	DIMENSIONS	SQ FT	U or FACTOR*	TEMP DIFF	SENSIBLE BTUH	TONS	LATENT BTUH
ROOF OR CEILING		90,000	0.110	20.00	198,000	16.50	
GLASS EAST		1,500		15.00	22,500	1.88	
(Solar) SOUTH		2,160		50.00	108,000	9.00	
WEST		1,500		15.00	22,500	1.88	
GLASS SOUTH		5,160	1.130	20.00	116,616	9.72	
(Conduction)						0.00	
WALLS E,W,N		9,040	0.167	20.00	30,194	2.52	
(Conduction)		2,880	0.167	50.00	24,048	2.00	
LIGHTING		90,000	4	3.416	1,229,760	102.48	
PEOPLE, NO.	(NO.)	400	250	250	100,000	8.33	100,000
OFF. EQUIPMENT	MOTORS (HP)	25	2,550		63,750	5.31	
COMPUTERS	(KW)	10	3,416		34,160	2.85	
KITCHEN						0.00	
VENTIL. AIR, Sens	1.43 AC/HR	30,000 CFM	1.08	20.00	648,000	54.00	
VENTIL. AIR, Lat.	1.43 AC/HR	30,000 CFM	0.010	4,840		0.00	1,452,000
TOTAL COOLING LOAD,			SENSIBLE		2,597,528	216	
BTUH = Sq Ft x U x Temp. Diff.			LATENT		1,552,000	121	
* U, CLTD, CLF, KW, HP, WD CLTD = (Temp diff) - (Daily Range + 14)/2			GRAND TOTAL		4,149,528	337	267*
BTU/YEAR = Sq Ft x U x Avg Temp. Diff. x Hrs							

* Sq Ft/Ton

Suburban Office Building

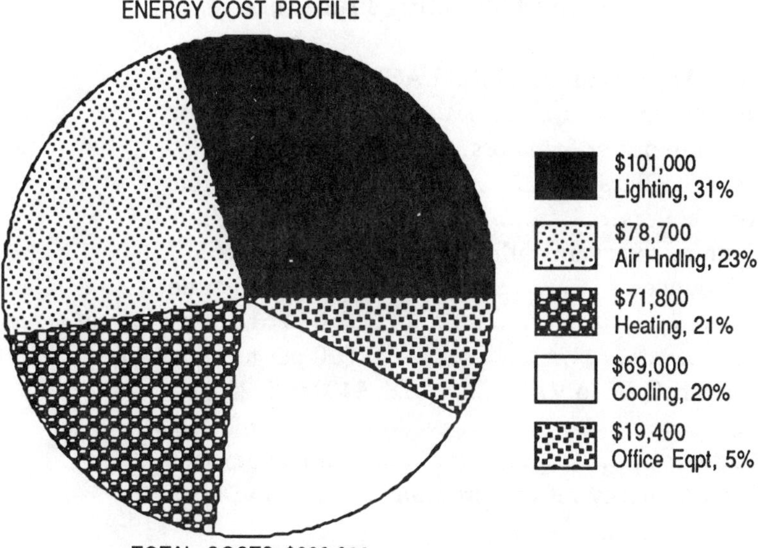
ENERGY COST PROFILE
$101,000
Lighting, 31%
$78,700
Air Hndlng, 23%
$71,800
Heating, 21%
$69,000
Cooling, 20%
$19,400
Office Eqpt, 5%
TOTAL COSTS $339,900

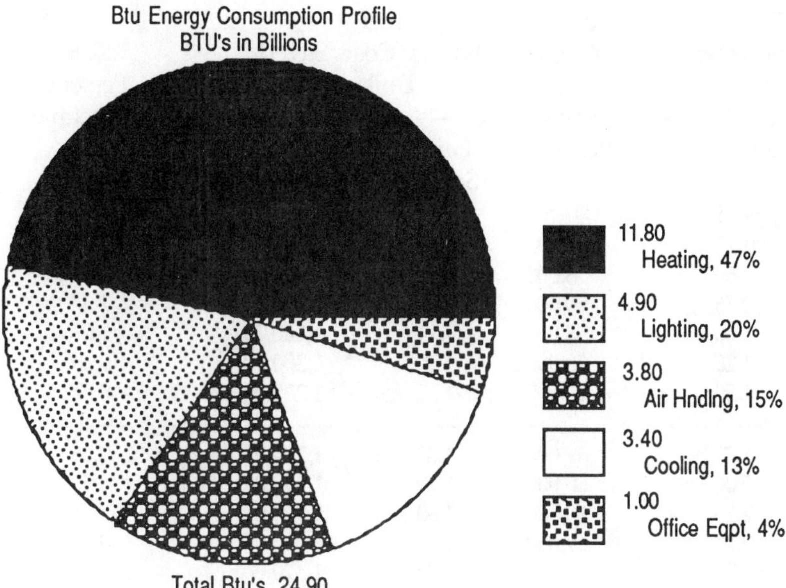
Btu Energy Consumption Profile
BTU's in Billions
11.80
Heating, 47%
4.90
Lighting, 20%
3.80
Air Hndlng, 15%
3.40
Cooling, 13%
1.00
Office Eqpt, 4%
Total Btu's, 24.90

Budget Estimating Energy Retrofit Costs and Estimating Procedures

COSTS TO REMOVE AND REPLACE

1. New pumps....................................$3 to $10/gpm
2. New fan driver & motors.............$100/hp+
3. New chillers...............................$200 to $300/ton
4. New boilers ....$6 to $300/ton
5. Heat recovery installations............$1 to $3/cfm
6. Rebalance....................................$10 to $20 outlet
   ....$60 to $120 per ahu coil
   ....$20 per terminal unit
7. Convert CAV to VAV..................$400 to $1200/box
   ....$.35 to $1/sq. ft
8. Audits ....2¢ to 5¢/sq. ft.
9. Overall energy retrofit programs......$.25 to $4/sq. ft.

Budget Estimating Energy Retrofit Costs Compared to Btu and Costs Savings

Retrofit Investment Cost Per Sq Ft Building	Rough Yearly Btu's Saved Per Sq Ft	Yearly Energy Costs Saved Dollars Per Sq Ft	Yearly Energy Costs Saved Dollars Per 1,000 Sq Ft	ROI Percent In Decimals
Office Buildings				
$0.25	40,000	$0.86	$860.00	3.37
0.50	46,000	0.99	990.00	1.91
0.75	50,000	1.08	1,080.00	1.37
1.00	53,000	1.14	1,140.00	1.07
1.25	56,000	1.20	1,200.00	0.89
1.50	58,000	1.25	1,250.00	0.77
1.75	59,000	1.27	1,270.00	0.66
2.00	60,000	1.29	1,290.00	0.58
2.25	60,500	1.30	1,300.00	.051
2.50	61,000	1.31	1,310.00	0.46
2.75	61,500	1.32	1,320.00	0.41
3.00	62,000	1.33	1,330.00	0.38

Hospitals				
$0.25	45,000	$0.97	$970.00	3.81
0.50	56,000	1.20	1,200.00	2.33
0.75	62,000	1.33	1,330.00	1.71
1.00	65,000	1.40	1,400.00	1.33
1.25	67,000	1.44	1,440.00	1.09
1.50	70,000	1.51	1,510.00	0.94
1.75	72,000	1.55	1,550.00	0.82
2.00	74,000	1.59	1,590.00	0.73
2.25	75,000	1.61	1,610.00	0.65
2.50	76,000	1.63	1,630.00	0.59
2.75	77,000	1.66	1,660.00	0.54
3.00	78,000	1.68	1,680.00	0.49
Universities				
$.025	30,000	$0.65	$650.00	2.53
0.50	52,000	1.12	1,120.00	2.17
0.75	61,000	1.31	1,310.00	1.68
1.00	67,000	1.44	1,440.00	1.37
1.25	72,000	1.55	1,550.00	1.17
1.50	77,000	1.66	1,660.00	1.04
1.75	80,000	1.72	1,720.00	0.92
2.00	84,000	1.81	1,810.00	0.84
2.25	87,000	1.87	1,870.00	0.76
2.50	90,000	1.94	1,940.00	0.71
2.75	92,000	1.98	1,980.00	0.65
3.00	93,000	2.00	2,000.00	0.60

1. Total savings and ROI based on electrical and fuel consumption.
2. Electrical savings based on 66 percent of total consumption.
3. Gas and oil based on 34 percent of total consumption.

Budget Estimating Chiller Retrofits

300-Ton Chiller

	Material Costs	Labor Hours	Totals w/30% MU
Chiller @ $200/ton	$60,000	130	$75,000
Piping	$1,000	100	$5,000
Electrical	$1,000	100	$5,000
Insulation	$1,000	100	$5,000
Controls	$1,000	100	$5,000
General Construction	$1,000	100	$5,000
Total	$65,000	100	$5,000

Hence $100,000 divided by 300 tons = $333 per ton

Estimating Retrofit Costs

Estimating is calculating labor, material and overhead costs on a project and coming up with a reasonably accurate total price which properly reflects the final actual costs of material and labor of the particular project.

Pricing is frequently based on imperfect or incomplete data, vague and inadequate plans and specs, and unpredictable and uncontrollable conditions creating an element of risk and potential error.

Hence, it is sometimes a relatively difficult, complicated and a judgement type affair and consequently, must be handled with great care using valid price charts, quotations, etc.

The procedure involves that the engineering design be completed first, either in complete detail, conceptually or practically designed in scope fashion.

A listing should be made of all types of items required.

Then quantities, types, sizes, accessories, building conditions, etc. must all be established and extended for labor and material. Sub-contractor and supplier quotations must be obtained and then all items summarized.

Converting Constant Volume Systems to VAV

CONSTANT VOLUME TERMINAL REHEAT CONVERSIONS

1. Constant volume terminal reheat systems are inefficient because they cool the main supply air first, in order to handle the zone requiring the most cooling, and then reheat it for other zones requiring less cooling or needing heat.
2. They can be converted to VAV by : a) Retrofitting the CAV terminals, b) By installing a VAV damper before the existing terminal, c) By replacing them with new VAV terminals.
3. Use a type fan volume control appropriate to the type and configuration of existing HVAC equipment, the system capacities, occupants' needs, existent static pressures, ductwork design etc. For larger, high and medium pressure type systems use motor frequency inverter, inlet van dampers etc. For smaller, lower pressure systems use discharge dampers as well as inlet dampers, ride fan curve, bypass ducts etc.

Procedure for Preparing a Retrofit Estimate

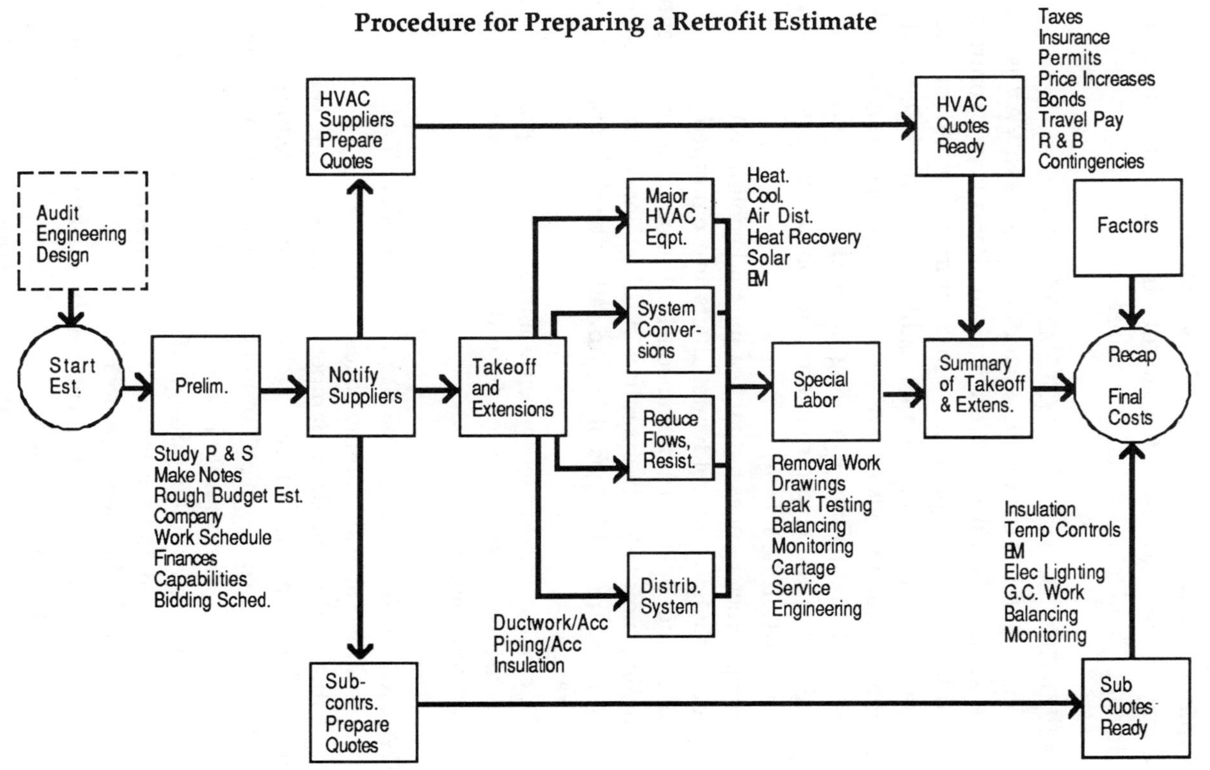

4. In order to monitor the variable demand of air at the terminals and to regulate the duct pressure and fan volume either a duct static pressure sensor is required in the main supply duct about two thirds of the way downstream from the fan, or air volume monitors in the main supply and return ducts are required.

CONVERTING DUAL DUCT SYSTEMS TO VAV

1. One method of converting dual duct systems to VAV is to cap the hot air inlets on the dual duct terminal and install controls on the cold inlet to vary flow between maximum and minimum. The hot deck dampers in the central air handling unit are closed and the operation of heating coils shut off. The room thermostat continues to control the dual-duct box, throttling the cold duct flow now instead of blending hot and cold air streams. Supply air reset is commonly added as part of the retrofit.
2. For perimeter zones a dual heating and cooling VAV system is created. Both hot and cold inlet ducts are equipped with VAV controls but not operated simultaneously. The hot deck in the central air handling unit is left in operation but the temperature of the hot air is reduced from old highs of 110° to 140° to a minimum, in the 80° to 100° range.
3. Duct static pressure sensor senses increase in duct pressure or air flow measuring station senses flow decrease and fan volume is adjusted accordingly.

LOW PRESSURE MULTIZONE SYSTEM CONVERSION

1. Install variable air volume boxes in each zone.
2. Either close hot deck dampers permanently in central air-handling unit and shut off flow to heating coils, or reduce hot deck temperature from old high of 110° to 140° F down to 80° to 90° range.
3. If hot deck temperature is lowered, or if heat is turned off and return air is recirculated through the hot deck, terminal goes to minimum first at master terminal, then directs cold deck at air handler to close while hot deck opens.
4. If hot deck is closed permanently operate as combination cooling, reheat terminal systems.
5. If building has separate perimeter heating and spaces are not under roof heating, close hot deck permanently.

Chapter 8
DESIGNING VAV SYSTEMS

The proper design of the air distribution system for VAV systems is extremely important and more complex than constant volume systems. Many more things must be taken into consideration and checked.

OVERALL VAV SYSTEM DESIGN PROCEDURE

HVAC Requirements and Zoning

1. Determine the occupant's HVAC requirements and design criteria.
   For example, general occupied areas may need to be 76°F at 35 percent RH for comfort, whereas another area may have to be 60°F at 50% RH for operational reasons.
2. Divide the building up into required zones of temperature control.
   Typical zoning might be separate east, west, south and north perimeter zones, and one or more separate interior zones.

Loads

3. Calculate heating and cooling loads for the overall building and for each zone at maximum and minimum conditions.
   Due to the shifting loads the east and west zones may require 14 tons of cooling and 5600 CFM each when they reach their peaks

during the day. They may only require 4 tons and 1600 CFM at minimum loads.

4. Calculate the amount of CFM needed for the various zones and for individual rooms.
5. Determine the peak simultaneous load if diversity is feasible.

Type System and Central Equipment

6. Select the most appropriate type of VAV system and central equipment to be used.
   A large high-rise office building normally required pressure independent terminals, variable fan volume, medium pressure duct systems, cooling the year around in interior zones and perimeter terminals with reheat terminals. Whereas, a small low-rise office building can live with pressure-dependent boxes, riding the fan curve rather than fan volume reduction, and perimeter baseboard heating.
7. Divide building up into systems.

Air Distribution Systems

8. Route and size ducts.
9. Select the types and size of terminal units.
10. Select the type and size of outlets.
11. Calculate duct system resistance for each system at maximum and minimum flows.
12. If there is diversity, determine if the supply fan will be sized at the diversity load or at total terminal CFM load.
13. Select fans and method of varying fan volume. Check performance at maximum and minimum and under diversity if included.
14. Select HVAC units.

Central Heating and Cooling Equipment

15. Select the type and size of central heating and cooling equipment needed, boilers, chillers, cooling towers, etc.

Hydronic Distribution System

16. Route and size piping.
17. Select the type and size of air handling unit coils and terminal coils.

18. Calculate the piping system resistance.
19. Select the type and size pump required.

Control System

20. Select and locate thermostats.
21. Select size and locate static pressure controller in main duct. Determine if really needed for small system.
22. Select air flow stations if used.
23. Select VAV control panel.
24. Route tubing, if a pneumatic control system, and select the type and size compressor along with appropriate valves and accessories.

VAV DUCT DESIGN

General Requirements

1. The pressure at the intake of the VAV terminal must be sufficient to operate the regulator plus overcome the resistance of the box discharge ductwork and end outlet.
2. Larger systems over 15,000 CFM total and with velocities over 2400 fpm should be designed with the static pressure regain method.
3. The equal friction method can be used for low velocity systems under 2400 fpm.
4. Duct runs must be sized for the maximum flow in each particular zone duct.

Design Criteria

The effect of maximum and minimum flows on the following items is critical.

a. Ventilation requirements.
b. Humidity
c. Proper air flow out of supply outlets
d. Performance of the VAV terminal
e. Noise levels
f. Back pressure in the ductwork
g. Pressure changes at the fans

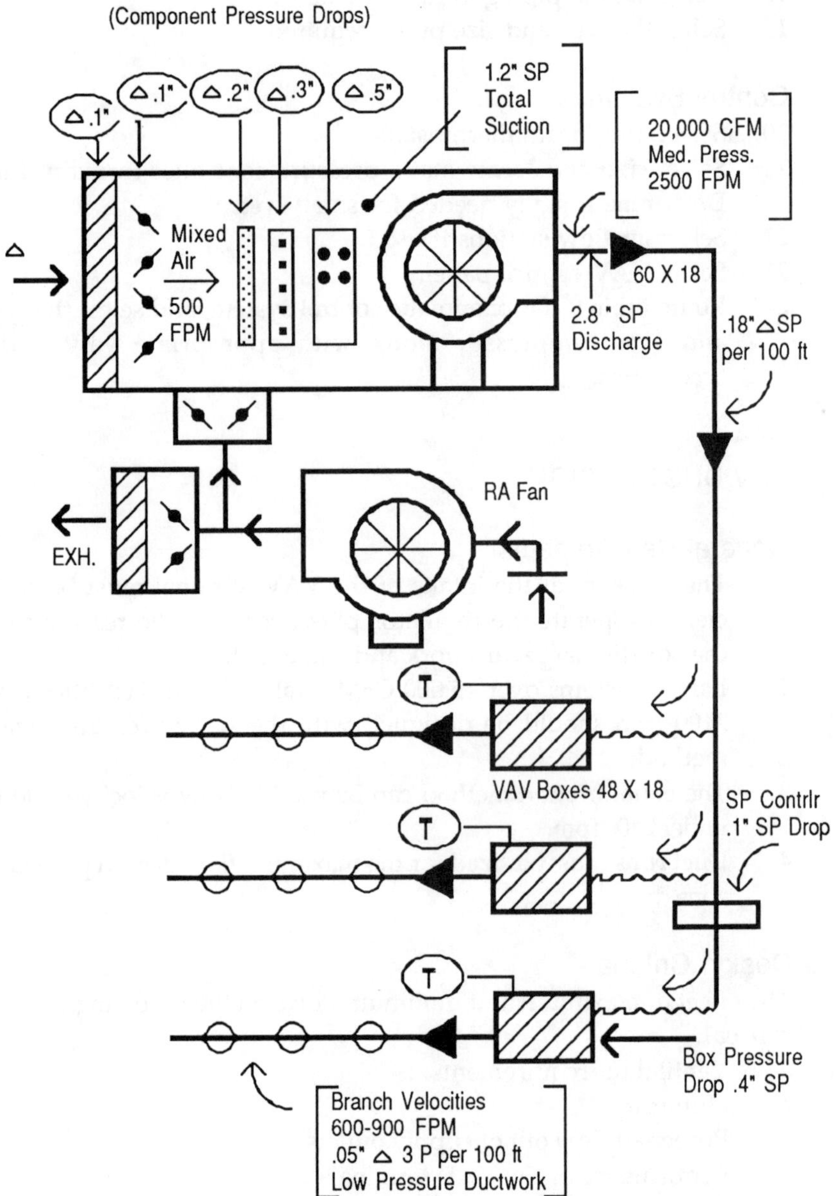

Figure 8-1. VAV Air Distribution Design

h. Fan performance
i. Proper flow and pressure drops in ductwork

Maximum Air Quantities

Zone ductwork and variable volume terminals must be sized according to the **governing factor**, either summer or winter sensible, latent, or ventilation loads.

The maximum fan CFM is equal to the sum of all peaks from governing zone latent or ventilation loads when the entire system reaches its peak sensible load.

When determining the total fan CFM in addition to meeting space heat and cooling loads, adjust air quantities for:

Leakage through outside dampers
Building exfiltration or infiltration
Building stack effect
Exhaust fans
Duct leakage
Duct heat gains or losses
Space pressurization

Minimum Air Quantities

Minimum air quantities in each zone and at the fan must both be examined in design. First, check for adequate dehumidification and ventilation. Second, check the resulting increase in back pressure at the terminal units.

If the increase in static pressure at the fan discharge and at the last terminal device at minimum system volume does not cause fan instability, distribution imbalance or excessive pressure drop in the last terminal, fan volume regulation and/or fan static pressure control may not be necessary.

Calculating Governing CFM's

The peak CFM load may be governed by sensible or latent room cooling loads, heating loads, outdoor air requirements, air motion, or exhaust.

When Supply Air for Cooling Governs:

Summer room supply CFM required to satisfy peak sensible load of each room $= \dfrac{\text{Peak summer sensible load for each room}}{1.1\ (TR - TS)}$

When Supply Air for Dehumidification Governs:

Summer room supply CFM required to satisfy peak sensible load of each room $= \dfrac{\text{Peak summer latent room load for each room}}{4840\ (WR - WS)}$

When Supply Air for Heating Governs:

Winter room supply CFM required to satisfy peak heating load to each room $= \dfrac{\text{Peak winter sensible room load (less any auxiliary heat), for each room}}{1.1\ (TS - TR)}$

When Outdoor Air CFM Governs:

Room supply air for ventilation requirements $= \dfrac{\text{Minimum outdoor air required in a room}}{\text{Ratio of systems total OA to total supply air}}$

When Exhaust Governs:

Room supply air for ventilation = Air exhausted or relieved from room and not returned to conditioned air system.

where:

TR = Temperature return air
TS = Supply air temperature required to satisfy the summer or winter peak loads
WR = Room humidity ratio
WS = Humidity ratio of dehumidified supply air

Minimum Ventilation Air Quantities can be maintained by:

Raising main supply air temperatures.

Raising reheat coil discharge temperatures.

Providing auxiliary heat in the room independent of the air system such as baseboard heating.

Providing individual zone fan recirculation and blending varying amounts of supply with room air or supply with ceiling plenum air.

Utilizing variable volume induction unit recirculation.

Throttling Ratio

The designer should distinguish between the minimum throttling ratio of any zone and that of the fan. The ratio can be a little at 0.25 in a zone, with a peak of 3 CFM/sq ft, without necessarily violating ventilation criteria, while the overall system ratio might never go below 0.5. The lower the system design static, the less need there is for fan volume regulation.

Diversity

If the peaks in each zone do not occur simultaneously, a diversity factor may be justified.

Diversities are mainly caused by a shifting sun load on the east, south and west exposures, outdoor temperature swings and ventilation loads, each having its peak load at different times during the day. The result is that the maximum simultaneous peak load may be 20 or 30 percent less than the sum of all the terminal units. A typical diversity situation is where the sum of all the terminal units is 32,000 CFM, but the fan is only sized for 24,000 CFM, because no more than that would even be needed simultaneously in any instant in time.

Return Plenum Ceilings

It is important to properly evaluate the percent of total light heat which is picked up by the return air in a return air hung ceiling. If at design cooling loads, room sensible heat has been reduced so much by recessed or return air lighting troffers that the design CFM approaches the minimum criteria, there will be no leeway for variable volume reduction at part loads.

Perimeter Air Systems

The perimeter type air system with a separate fan may be designed with 100 percent recirculation using no outdoor air. This takes advantage of all light, heat, etc. in the portion of the return air fed back to the perimeter system and fulfills the heating requirements during a substantial portion of the heating season. When an internal heat recovery system is used with a heat pump, all internal heat can be applied to perimeter heating.

OUTLET SELECTION

For best results diffusers should be selected for full volume, maximum permissible outlet velocity, and correct throws, to have maximum leeway for reduction in flow.

There are, however, several important considerations for VAV air distribution using outlets for **horizontal discharge patterns.**

1. **An outlet with a low throw coefficient** should be used. A small throw coefficient gives a smaller absolute change in the throw values with variation in volume and thus tends to minimize the change in air motion within the occupied space.
2. Outlets should be chosen for **small quantities of air**. In this manner absolute values of throw will vary a minimum with the in variation flow rate for the outlet. It is recommended that round diffusers should not exceed 300 to 350 CFM.
3. **For under window types of distribution**, vertical throw outlets with nonspreading pattern should be used. To prevent cool air dropping back into the occupied space at minimum flow conditions, the outlet discharge velocity should be 500 fpm minimum. Throw coefficient would be higher to project air up to the ceiling.
4. **For small volume reduction requirements**, all types of supply outlets have a tolerance in throw and drop that permits their use without concern. Volume reductions for round or square diffusers should exceed the 25 to 35 percent range.
5. **Modular ceiling slot-type diffusers** are considered to have the most desirable characteristics for large volume reductions. They should be located so throws do not exceed 150 fpm. Impinging velocities between diffusers should not exceed 100 fpm.

Plenum slot diffusers can handle variations in air flow that approach zero without dumping and still maintaining horizontal air flows.

6. **Air flows** in the spaces at occupied levels should range between 10 fpm and 45 fpm for comfort. Under 10 fpm causes affecting of stagnation and over 45 fpm causes a draft feeling.

Sound Levels

Air distribution is very important in variable volume applications. Consideration must be given to distribution and to sound levels at maximum and minimum flow. If the combined sound level of the terminal unit and diffuser at maximum flow is at least 3 dp below the room ambient sound level, variations will not be noticed.

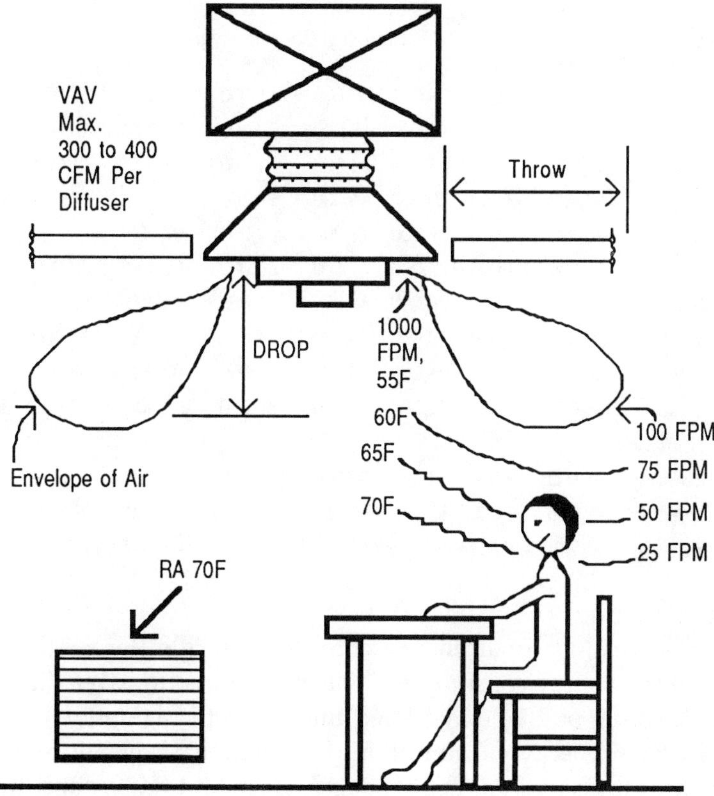

Figure 8-2

CALCULATING SEASONAL ENERGY CONSUMPTION

The range for the rate of seasonal heat loss for winter of a building can vary astronomically. Office buildings in the U.S. may be as low as 20,000 Btu/sq st/Yr and as high as 150,000 Btu/sq ft/Yr and may average 30,000.

Hence, a 90,000-sq-ft building may have a total seasonal heat loss of 1.8 billion Btu in a well insulated building in the southern part of the U.S., while be six times as much–10.8 billion Btu–in a leaky, poorly insulated, single-story building in the north.

Average Annual Energy Performance in Btu's Per Square Feet

Building Type	National	Heating and Cooling Degree Day Region* 1	2	3	4	5	6	7
Office	84,000	85,000	76,000	65,000	61,000	51,000	50,000	64,000
Elementary	85,000	114,000	70,000	68,000	70,000	53,000	48,000	57,000
Secondary	52,000	77,000	65,000	55,000	51,000	37,000	41,000	34,000
College/Univ.	65,000	67,000	70,000	46,000	59,000			83,000
Hospital	190,000		209,000	171,000	227,000	207,000		197,000
Clinic	69,000	84,000	72,000	71,000	65,000	61,000	59,000	59,000
Assembly	61,000	58,000	76,000	68,000	51,000	44,000	68,000	57,000
Restaurant	159,000	162,000	178,000	186,000	144,000	123,000	137,000	137,000
Mercantile	84,000	99,000	98,000	86,000	81,000	67,000	83,000	80,000
Warehouse	65,000	75,000	82,000	65,000	50,000	38,000	37,000	39,000
Residential Non-Housekeeping	95,000	99,000	84,000	94,000	125,000	90,000	93,000	106,000
High Rise Apt.	49,000	53,000	53,000	52,000	53,000	84,000	20,000	

***See Auditing Procedures, Chapter 3, pg. 000** February 1978, HUD-PDR-290

Total Seasonal

The total seasonal energy consumption is different than the peak hourly energy consumption which is figured to cover the worst possible condition that may be encountered in the winter or summer. The seasonal energy consumption deals with averages of temperatures, heat transmission factors, degree days and time spans. It also takes into consideration the heat gains generated in the building by lights, people, etc. which is not a factor in the peak heat loss or gain.

ENERGY CONSUMPTION OF BUILDING DEPENDENT ON:

1. R values of walls, windows, roofs, doors and resultant "U" factors.
2. Percentage of glass and R factors.
3. Amount of outside air being brought in either by the supply fan or by infiltration or both.
4. Amount of air being exhausted.
5. Size of building.
6. Ratio of building skin surface to floor square footage.
7. Solar orientation.
8. Average outside and indoor temperature. Average outdoor and indoor humidity levels.
9. Occupancy levels and operational times.
10. Occupancy levels and operational times.
11. Efficiency of equipment and degree of maintenance.
12. Internal heat gain due to lighting, occupancy, equipment and solar heat.

RATIO OF SKIN TO FLOOR AREA FACTOR

The more outside skin surface there is in relationship to sq. ft. of floor area, such as with a single-story building, the greater the rate of energy consumption. The smaller the ratio, as with multi-story buildings, the lesser the rate of energy consumption.

For example:

A single-story building with a ratio of 1.5 skin surface to floor surface area. Rate of heat losses or gains may be 50 to 100% greater than multi-story buildings.

A two-story building with a 1 to 1 ratio. Energy consumption may be 25% less than single story.

Multi-story buildings with .5 to one ratio. Rate of energy consumption may be 50% less than single-story.

OVERALL BUILDING SKIN FACTOR

The **yearly skin loss formula** is (does not include OA and infiltration losses nor building heat gain):

Btu/Yr = (Degree Days) x (Hours/day) x (U Factor Average) x (Sq Ft Skin)

Example: 90,000-sq.-ft., single-story building, 14 ft. high.

Btu/Yr = 6600 x 24 x .2 average U x 107,080 sq. ft.
= 3.393 Bill Btu (average 37,692 Btu/sq ft/yr)

WINTER INTERNAL HEAT GAINS

The annual heat loss of a building structure is reduced by internal heat gains in the winter as follows:

1. **Lighting heat pickup** which may constitute a 25 to 50 percent regain. At 3 watts per sq ft.

   3W x 3.416 Btu/W x 90,000 sq. ft. x 2500 hr. = 2.31 Bill Btu/Yr

2. **People heat pick-up** at occupancy rate of 1 person per 225 sq. ft.

   $$\frac{90{,}000 \text{ sq ft}}{225 \text{ sq ft/person}} = 400 \text{ people}$$

   400 People x 250 BtuH x 2000 Hr = 200 Mill Btu/Yr

3. **Solar radiation gains** in winter of glass, roof, walls.

4. **Equipment heat gains:**

1 HP = 2550 BtuH

(25 HP) x (2550 BtuH) x (2000 Hrs) = 127 Mill Btu/Yr

If the infiltration for a 90,000-sq-ft office building is one half an air change per day, the seasonal heat loss is calculated as follows:

$$\text{CFM} = \frac{\text{Air Changes/hr x cu ft of bldg}}{60}$$

$$= \frac{.5 \times 630{,}000}{60} = 10{,}500 \text{ CFM}$$

Btu/yr = CFM x 1.08 x TD x hours

Btu/yr = 10,500 x 1.08 x 30 x 4800 hr = 1.633 Bill Btu

QUICK APPROXIMATION OF INSTANTANEOUS HEAT LOSSES

A quick, approximate method of determining the overall heat loss of a building is to read the gas meter and relate to outside air temperature as follows:

1. Read gas meter or oil gpm meter.
2. Add calculated lighting, people, solar and equipment heat gains if done during occupancy.
3. Subtract inefficiency of heating apparatus.
4. Relate to concurrent outside temp.

Example: One Hour Gas Meter Reading
Average Outside Air 35°F:
90,000-sq-ft building
3 Watts per sq ft
400 people

1. Reading: 2000 cu ft of gas in 1 hr = 2 mill BtuH

2. Calculate heat gains:

   a) 3 watts per sq ft x 90,000 sq ft = 270,000 watts
   270,000 W x 3.413 = 92,150 BtuH

   b) Occupancy, 400 people x 250 = 100,000 BtuH

   c) Take reading in darkness to eliminate calculations of solar gain.

   d) Building equipment: 45 kWh x 3413 = 153,600 BtuH
   Total Heat Gain = 345,750

3. Divide by efficiency of heating equipment

$$\frac{345{,}750}{.75} = 461{,}000 \text{ input}$$

4. Actual Building BtuH Heat Loss at 35°F OA = 2,000,000
   + 461,000
   Total 2,461,000

5. It it's possible to read gas or oil meter at night during non-occupancy with all lighting and equipment turned off, total vacancy and with no solar load, the meter reading will give a true representation of the building heat loss if internal heat gains from HVAC equipment is disregarded.

Sample
HEAT LOSS CALCULATION

☐ PEAK PER HR ☒ SEASONAL ☒ EXISTING ☐ NEW

Building SUBURBAN OFFICE BUILDING Date

Location Latitude 41

Type Building OFFICE AND LABS Stories 1 When Built 1967

Sq Ft Area 90,000 Cubic Ft of Space 1,260,000

X Calculation for Whole Building For Partial Area

Budget Load: Sq Ft 90,000 x BTUYR/Sq Ft 78,500 = 7,065 Mill BTUYR

Outside Design: DB -10 WB Avg OA Temp. 35 Winter Hours 4,800

Inside Design: DB(day) 74 RH DB(night)

ITEM	DIMENSIONS	SQ FT	U	TEMP DIFF	BTU: Per Hour Seasonal	SEASONAL HOURS
ROOF OR CEILING	360 X 250	90,000	0.110	39.00	1,853,280,000	4,800
FLOOR	360 X 250	90,000	0.100	16.00	691,200,000	4,800
GLASS	950 X 7	6,650	1.100	39.00	1,369,368,000	4,800
DOORS						
WALLS	1220 X 14	10,430	0.330	39.00	644,323,680	4,800
COLD INSIDE WALLS						
VENTILATION	1.43 AC/HR CFM=	30,000	1.080	39.00	6,065,280,000	4,800
DUCT LOSSES						
GROSS TOTAL HEAT LOSS					10,623,451,680	
HEAT GAINS; LITES	4W/SQ FT	90,000	4.000	3.416	3,074,400,000	2,500
PEOPLE	No. People =	400	250		200,000,000	2,000
OFF. EQUIPMENT	HP =	25	2,550		127,500,000	2,000
COMPUTERS	KW =	5		3416.00	149,620,800	8,760
TOTAL INTERNAL GAINS					3,551,520,800	
NET TOTAL BUILDING HEAT LOSS					7,071,930,880	78,577*
INPUT TO HEATING EQUIPMENT, efficiency =			.60		11,800,000,000	131,110*

BTUH = Sq Ft x U x Temp. Diff. BTU/YEAR = Sq Ft x U x Avg Temp. Diff. x Winter Hours

* BTU/SQ FT/YEAR

Remarks____

Budget Estimating Peak Heating, Cooling and CFM Loads (Per Square Foot of Building)

Type Building		Cooling Btu Per Sq Ft	Load Sq Ft Per Ton	Heating Load Btu/ft	Supply CFM Per Sq Ft	Duct WT lbs/Per Sq Ft
Apartments, Condominiums		25	480	26	.8	2
Auditoriums		40	300	40	1.3	.8
Banks		48	250	26	1.6	1.2
Bowling Alleys		40	300	40	1.3	.6
Churches		36	330	36	1.2	.5
Clubhouses		50	240	40	1.6	1.0
Cocktail Lounges		70	170	30	1.3	1.2
Computer Rooms		140	85	20	4.5	1.5
Colleges:	Admin., Classrooms	44	270	42	1.5	1.3
	Dormitories	--	--	25	--	.3
	Gyms, Fieldhouses	--	--	40	.5	.3
	Science Bldgs.	54	220	55	1.8	1.4
Courthouses		50	240	35	1.6	1.2
Fire Stations		--	--	35	--	.7
Funeral Homes		30	400	32	.9	.8
Hospitals		44	170	28	1.4	1.5
Hotels		34	350	28	1.1	.5
Housing for Elderly		--	--	28	--	.2
Jails		25	480	30	.8	1.2
Laboratories		60	200	50	2.0	1.5
Libraries		46	260	37	1.5	1.3
Manufacturing Plants		40	300	36	1.3	.4
Medical Centers, Clinics		35	340	35	1.1	1.0
Motels		30	400	30	1.0	.5
Municipal Bldgs., Town Halls		45	265	36	1.4	1.2
Museums		34	350	40	1.1	.8
Nursing Homes		43	280	34	1.4	.5
Office Bldgs:	Low Rise	35	340	35	1.2	.5 to .8
	High Rise	40	300	30	1.3	1.1
Police Stations		42	285	35	1.3	1.0
Post Offices		44	270	40	1.4	1.2
Project Homes		--	--	28	--	.1
Restaurants		80	150	35	2.0	1.0
Residences		25	500	30*	.8	.3
Schools:	Elementary	--	--	26	.8	.7
	Middle, Jr. Highs	36	333	26	1.0	1.1
	High Schools	33	360	40	1.1	1.3
	Vocational	20	600	40	.9	1.5
Stores:	Beauty Shops	63	190	30	2.0	1.3
	Department	34	350	30	1.1	.7
	Discount	34	350	32	1.1	.4
	Retail Shops	50	340	32	1.6	.8
	Shopping Centers	30	400	30	1.1	.4
	Supermarkets	30	400	32	1.0	.4
Theatres		40	300	40	1.3	.8
Warehouses		--	--	20	--	.2

1. Cooling load based on 15° temp. difference, 50% RH, 400 CFM per ton
2. Heating load based on 70° temp. difference and is the output Btu. Apply inefficiency factor to heating equipment for input Btu
3. Price includes all HVAC material, labor and subs and overhead and profit

*R-11 wall and R-19 ceiling insulation.

Sample
COOLING LOAD CALCULATION

☒ PEAK PER HR ☐ SEASONAL ☒ EXISTING ☐ NEW

Building SUBURBAN OFFICE BUILDING Date

Location Latitude 41 Peak Load, HR, MO:

Type Building OFFICE AND LABS Stories 1 When Built 1967

Sq Ft Area 90,000 Cubic Ft of Space 1,260,000

X Calculation for Whole Building For Partial Area

Building Load: Sq Ft 90,000 x Sq Ft/Ton 270 = 333 Tons

Outside Design: DB 95 WB 75 Avg OA Temp. 35 Summer Hours 900

Inside Design: DB(day) 74 RH DB(night) 74

ITEM (Orient.)	DIMENSIONS	SQ FT	U or FACTOR*	TEMP DIFF	SENSIBLE BTUH	TONS	LATENT BTUH
ROOF OR CEILING		90,000	0.110	20.00	198,000	16.50	
GLASS EAST		1,500		15.00	22,500	1.88	
(Solar) SOUTH		2,160		50.00	108,000	9.00	
WEST		1,500		15.00	22,500	1.88	
GLASS SOUTH		5,160	1.130	20.00	116,616	9.72	
(Conduction)						0.00	
WALLS E,W,N		9,040	0.167	20.00	30,194	2.52	
(Conduction)		2,880	0.167	50.00	24,048	2.00	
LIGHTING		90,000	4	3.416	1,229,760	102.48	
PEOPLE, NO.	(NO.)	400	250	250	100,000	8.33	100,000
OFF. EQUIPMENT	MOTORS (HP)	25	2,550		63,750	5.31	
COMPUTERS	(KW)	10	3,416		34,160	2.85	
KITCHEN						0.00	
VENTIL. AIR, Sens	1.43 AC/HR	30,000 CFM	1.08	20.00	648,000	54.00	
VENTIL. AIR, Lat.	1.43 AC/HR	30,000 CFM	0.010	4,840		0.00	1,452,000
TOTAL COOLING LOAD,			SENSIBLE		2,597,528	216	
			LATENT		1,552,000	121	
			GRAND TOTAL		4,149,528	337	267*

BTUH = Sq Ft x U x Temp. Diff.
* U, CLTD, CLF, KW, HP, WD
CLTD = (Temp diff) - (Daily Range + 14)/2
BTU/YEAR = Sq Ft x U x Avg Temp. Diff. x Hrs

* Sq Ft/Ton

Weather Data

Average Seasonal Conditions

City	State	Winter Avg DB Winter Temp	Winter Length in Weeks	Summer Avg DB Summer Temp	Summer Length in Weeks
Atlanta	GA	41.1	19.8	78.7	30.0
Boston	MA	35.1	31.1	76.0	19.8
Charleston	SC	43.3	14.2	78.7	36.0
Chicago	IL	34.2	30.0	77.0	20.9
Cleveland	OH	34.0	29.4	76.5	21.0
Dallas	TX	42.5	15.1	82.8	34.6
Denver	CO	35.2	29.4	77.9	22.6
Detroit	MI	33.8	30.5	75.8	19.2
Houston	TX	47.0	6.0	80.3	42.0
Indianapolis	IN	35.8	26.7	78.0	23.9
Las Vegas	NV	43.7	15.6	86.8	35.4
Miami	FL	49.3	1.6	80.4	50.1
Milwaukee	WI	33.0	30.0	77.0	20.9
Minneapolis	MN	29.3	31.0	76.8	18.8
Nashville	TN	39.3	23.3	79.7	28.4
New Orleans	LA	46.4	9.4	79.8	39.6
New York City	NY	38.0	27.5	76.0	20.0
Philadelphia	PA	38.2	26.0	77.5	23.7
Portland	OR	44.0	30.8	73.5	15.8
Richmond	VA	40.9	20.9	77.8	26.8
Salt Lake City	UT	36.5	30.2	79.0	19.9
Seattle	WA	43.7	37.3	70.9	9.4
St Louis	MO	36.1	24.2	79.6	26.3
Trenton	NJ	37.5	26.9	77.1	22.9

Minimum U's or R's for Walls for Heating

Heating Degree Days	Minimum Acceptable U	R
1000	.40	2.5
2000	.30	3.33
3000	.30	3.33
4000	.20	5.0
5000	.20	5.0
6000	.15	7.0
7000	.15	7.0
8000	.10	10.0
9000	.10	10.0

Minimum U's or R's for Roofs for Heating

Heating Degree Days	Minimum Acceptable U	R
1000	.30	3.3
2000	.20	5.0
3000	.20	5.0
4000	.15	7.0
5000	.15	7.0
6000	.10	10.0
7000	.10	10.0
8000	.06	17.0

Rates of Heat Gain from Occupants of Conditioned Spaces

Degree of Activity	Typical Application	Total Heat Adults, Male Btu/h	Total Heat Adjusted[b] Btu/h	Sensible Heat Btu/h	Latent Heat Btu/h
Seated at rest	Theater, movie	400	350	210	140
Seated, very light writing	Offices, hotels, apts.	480	420	230	190
Seated, eating	Restaurant[c]	520	580	255	325
Seated, light work, typing	Offices, hotels, apts.	640	510	255	255
Standing, light work or walking slowly	Retail Store, bank	800	640	315	325
Light bench work	Factory	880	780	345	435
Walking, 3 mph, light machine work	Factory	1040	1040	345	695
Bowling[d]	Bowling alley	1200	960	345	615
Moderate dancing	Dance hall	1360	1280	405	875
Heavy work, heavy machine work, lifting	Factory	1600	1600	565	1035
Heavy work, athletics	Gymnasium	2000	1800	635	1165

R and U Values of Roofing and Wall Materials

Description	Density lb/cu ft	Thickness	R Per Inch Thickness	Resistance R for thick. shown	Conduct. U for thick. shown
AIR					
Outside air film (15 mph wind).			---	.17	5.88
Inside air film (still air)			1.64	.61	
Air space, horiz. flow		1/4"	---	.91	1.10
(50°F mean, 30 TD)		1/2'	---	1.23	.81
		3-1/2'	1.00	1.25	.80
Ceiling air space vert flow				1.00	

R and U Values of Roofing and Wall Materials (Cont'd)

Description	Density lb/cu ft	Thick-ness	R Per Inch Thickness	Resistance R for thick. shown	Conduct. U for thick. shown
MASONRY					
Concrete, med. wt	140.0	1"	.83	.83	1.20
Concrete, light wt	40.0	1"	.83	.83	1.20
Common brick	120.0	4"	.20	.79	1.27
	120.0	8"	.20	1.59	.63
Face brick	125.0	4"	.11	.43	2.33
Concrete block. L.W.	38.0	4"	.38	1.51	.66
	38.0	8"	.38	.71	1.41
Concrete block	61.0	4"	.50	2.02	.50
	61.0	8"	.50	4.04	.25
Clay tile	70.0	4"	.25	1.01	.99
	70.0	8"	.25	2.02	.50
INSULATION					
– Fiberglass	.75	1"	3.14	3.14	.32
	.75	6"	3.14	11.00	.09
	.75	6"	3.14	19.00	.005
Board	5.7	1"	3.33	3.33	.30
	5.7	2"	3.33	6.68	.15
	5.7	3"	3.33	10.00	.10
Cellular glass		4"	2.9	11.6	.09
Poly Styrene		4"	4.5	15.0	.06
Polyurethene		4"	6.25	25.0	.04
WOOD – Hardwoods	45.0	1"	.91	.91	1.10
Fir, Pine	37.0	1"	1.19	1.19	.84
	37.0	2"	1.19	2.39	.42
	37.0	2.5"	1.19	2.98	.34
	37.0	3"	1.19	3.58	.28
	37.0	4"	1.19	4.76	.21
Plywood		5/8"	1.25	.78	1.28
		1"	1.25	1.25	.80
Sheathing		5/8"	2.22	1.39	.72
		1"	2.22	2.22	.45
GLASS – Single Pane			---	1.10	.91
Double pane, 1" space			---	.52	1.92
FINISHING MATERIALS, WALLS, CEILINGS					
Plaster, Gypsum, Foil	100.0	5/8"	.95	.59	1.69
	100.0	3/4"	.95	.71	1.69
Plaster		3/4"			
Acoustic Tile		3/4"	2.34	1.79	.56
METAL – Metal Deck		20 Ga.	---		
ROOFING – Built up		3/8"	---	.33	3.03

Seasonal Energy Cost Formulas

Heating Cost Per Yr of Fuel For Transmission Losses

$$\text{Fuel Costs/Yr} + \frac{U \times A \times TD \times HR \times \$MMB}{1{,}000{,}000 \times effc}$$

Heating Costs Per Yr of Fuel for Heating Outside Air

$$\text{Fuel Costs/Yr} + \frac{CFM \times TD \times 1.08 \times HR \times \$MMB}{1{,}000{,}000 \times effc}$$

where:

U = Btu/F/Sq Ft, heat transmission
A = Area in sq ft
TD = Temperature difference between inside and outside surfaces
HR = Hours per season or year
$MMB = Cost per million Btu of fuel
effc = Efficiency of combustion

BASIS

1. 4800 seasonal hours typical for northern state winters.
2. 2600 seasonal hours typical for five 10-hour days of operation
3. 35°F typical average winter temperature for northern states and 70°F average inside temperature, hence the temperature difference is 35°F.
4. Based on natural gas fuel at $5 per million Btu which equals 50¢ per therm.

Fuel Heating Values

Fuel	**Heating Value**
Coal	
anthracite	13,900 Btu/lb
bituminous	14,000 Btu/lb
sub-bituminous	12,600 Btu/lb
lignite	11,000 Btu/lb

Heavy Fuel Oils and Middle Distillates

kerosene	134,000 Btu/gallon
No. 2 burner fuel oil	140,000 Btu/gallon
No. 4 heavy fuel oil	144,000 Btu/gallon
No. 5 heavy fuel oil	150,000 Btu/gallon
No. 6 heavy fuel oil, 2.7% sulfur	152,000 Btu/gallon
No. 6 heavy fuel oil, 0.3% sulfur	143,800 Btu/gallon
Gas	
natural	1,000 Btu/cu ft
liquefied butane	103,300 Btu/gallon
liquefied propane	91,600 Btu/gallon

Source: *Brick and Clay Record*, October 1972; reprinted with permission of the Cahner's Publishing Co., Chicago, Ill.

FORMULA FOR PERCENT OF ACCUMULATED ENERGY SAVINGS

The actual total savings is not the arithmetical sum of all the percentages saved on the individual energy conservation items. It is less than the sum of individual savings and the following computation must be made working with percentages.

1.

$$\text{Total Percent Savings} = 100\left[1 - (1 - SAV_1) \times (1 - SAV_2) \times (1 - SAV_3)\ldots\text{Etc}\right]$$

where: SAV_1 is percent of saving for each item.

2. Example problem, turn thermostats down and up:

	Now Setting	Savings
Heating	70%F	6%
Cooling	78%F	15%
Night setback	10%	7%
Total Percent	= 100 [1 – (1– .06) x (1 – .15) x (1 – .07)]	
Savings	= 26%	

COST PER MILLION BTU

Costs Comparisons for Different Fuels

There is no inefficiency loss in electrical heating. The input to an electrical heating coil is equal to the actual space heat loss. Gas and oil do have combustion efficiency losses, however, and the building heat loss calculations must be increased for combustion losses, in order to make a valid comparison to electrical heating costs.

1. GAS: The typical efficiency of natural gas is 75 percent, hence with input gas at $5.00 per million Btu (50¢ per therm):

$$\text{Real Cost Per Mill Btu Output} = \frac{\text{1 Mill Btu}}{.75} = \frac{\$5.00}{.75} = \$6.67$$

2. OIL: The typical efficiency of oil is 70 percent, hence with input oil at $7.00 per mill Btu (7 gal. at 143,000 Btu per gal.):]

$$\text{Real Cost Per Mill Btu Output} = \frac{\text{1 Mill Btu}}{.70} = \frac{\$7.00}{.70} = \$10.00$$

3. ELECTRICITY: Since there is no inefficiency factor involved with electric heat, the cost of input equals output. Then at 7¢ per kWh and 3416 Btu per kWh:

$$\text{Cost Per Mill Btu Output} = \frac{\text{1 Mill Btu}}{3416} \times 7¢ = 293\text{ kW} \times 7¢c = \$20.46$$

Hence the ratio electric heat runs to gas typically is:

Electric to gas: 3 to 1 ratio

Electric to oil: 2 to 1 ratio

HEATING DEGREE DAYS

Heating degree days are the number of degrees by which the mean temperature for a day falls short of 65°F. If, for example, the mean temperature on a given day were 45°F, that day contributed 20 degree days to the heating season.

The 65°F is assumed to be the actual design temperature required based on the assumption that internal heat gains, such as light, people, equipment, etc. will actually raise the indoor temperature to 70°F or 72°F.

This approach is primarily only accurate for residential heating load calculations, but the degree days are useful for ascertaining the severity of summers and winters, and for determining unknown actual number of winter days or hours and average temperature.

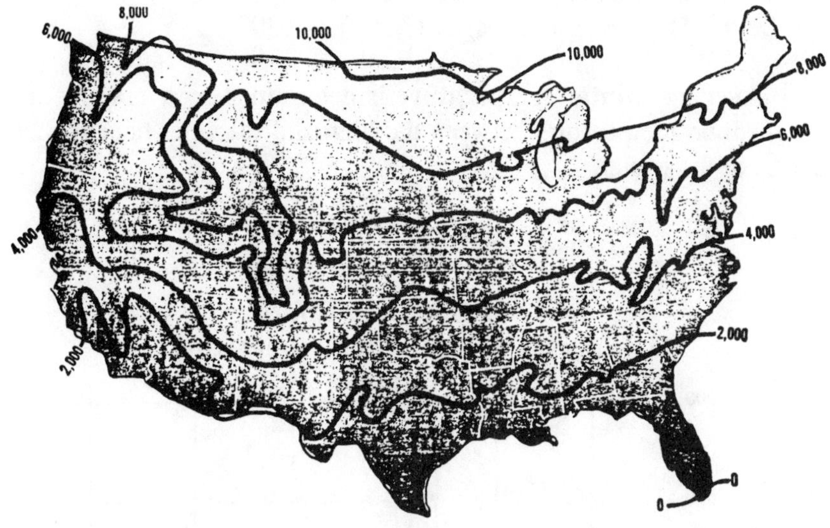

Figure 8-3

To determine the number of unknown winter hours from the given known degree days and average temperature:

If the average outdoor temperature for a particular winter in the Chicago area were 35°F and the given degree days 6000:

Average Temperature Difference = 65 – 35 = 30 TD

$$\text{Number of Winter Hours} = \frac{\text{Degree Days}}{\text{Average TD}} = \frac{6000}{30} \times 24 = 4800$$

To determine the average temperature from given, know degree days and number of winter days.

If number of winter days is 210 and the degree days are 6600:

$$\text{Average Temperature} = \frac{6600}{210} = 32°\text{F}$$

To determine the degree days from the given number of winter days and average temperature:

A Midwest city along the 41st parallel may have 210 days of winter that are below 65°F at an average outdoor temperature of 35°F.
Temperature Difference = 65 – 35 = 30
Degree Days = (210 days) x (30TD) = 6300

Average winter temperature is determined by averaging out the medium temperature between the high and lows per hour for the course of the winter.

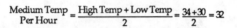

$$\text{Medium Temp Per Hour} = \frac{\text{High Temp} + \text{Low Temp}}{2} = \frac{34+30}{2} = 32$$

Annual Occupied Hours At Different Outside Air Temperatures

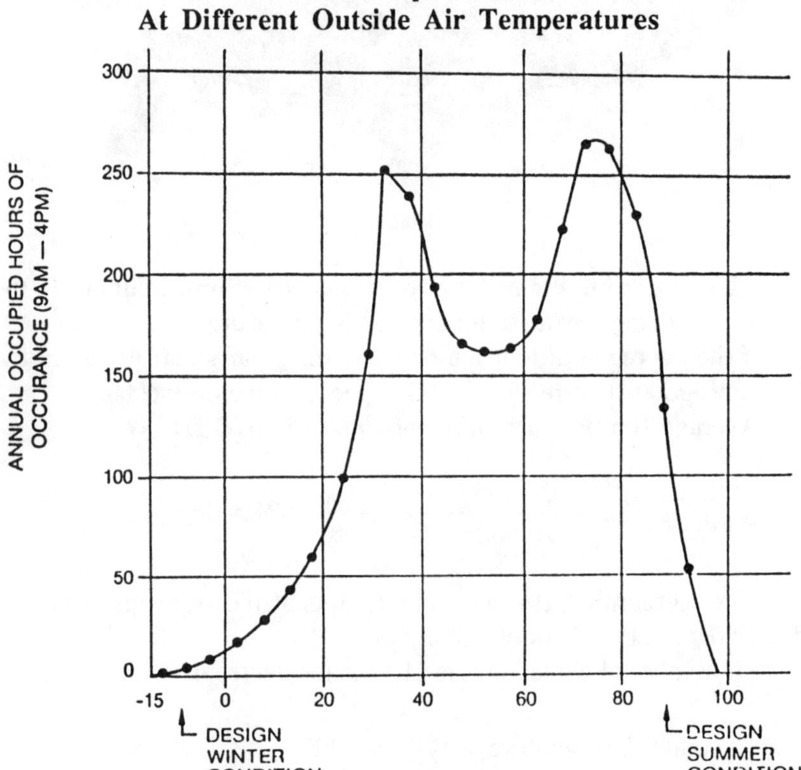

Chicago, Illinois – Bin weather data for a typical year expressed as a curve. As with most locations, peak outdoor conditions are the exception, not the rule.

Figure 8-4

Estimated Equivalent Rated Load Hours for a Normal Cooling Season

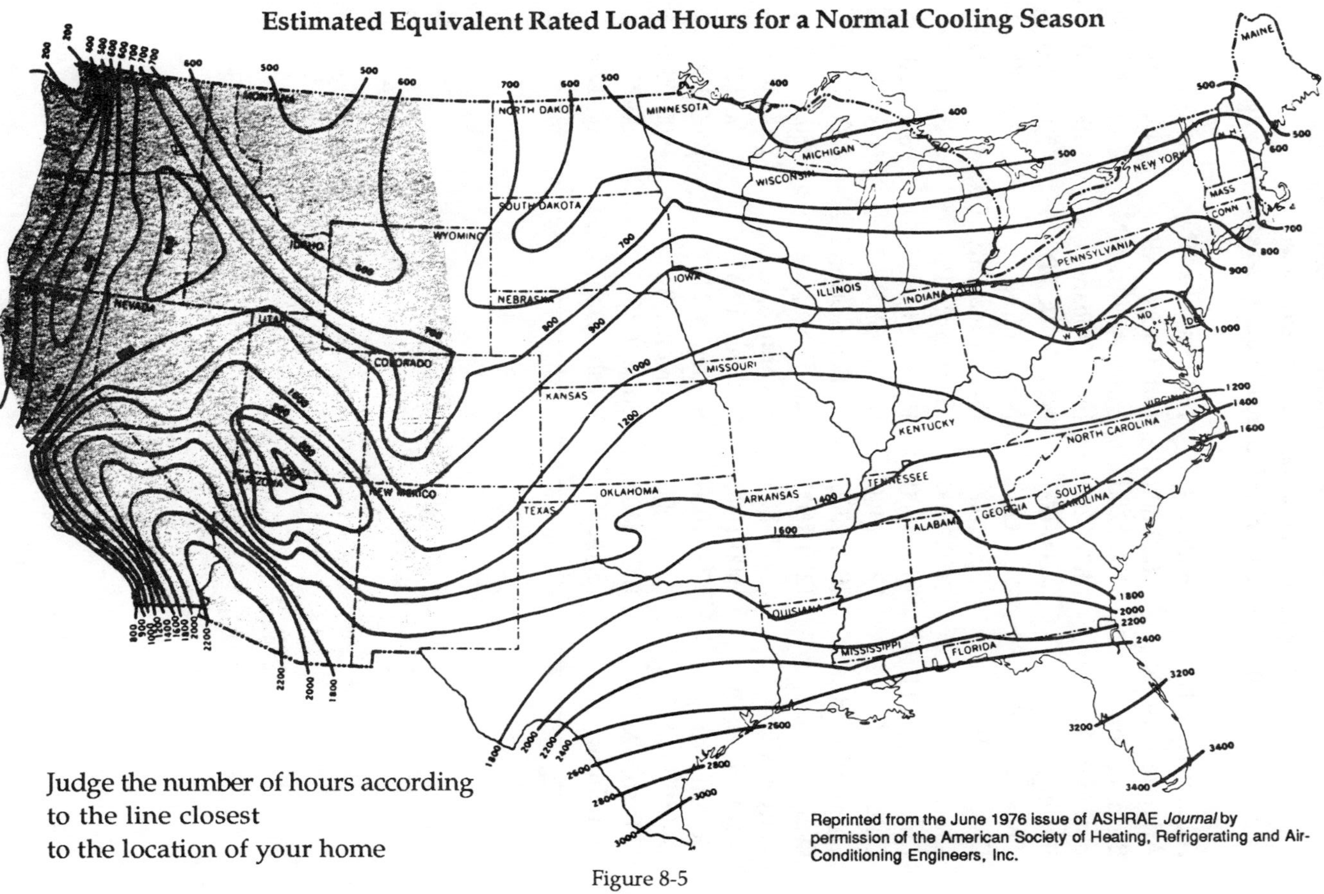

Judge the number of hours according
to the line closest
to the location of your home

Reprinted from the June 1976 issue of ASHRAE *Journal* by permission of the American Society of Heating, Refrigerating and Air-Conditioning Engineers, Inc.

Figure 8-5

Estimated Equivalent Rated Full Load Hours of Operation for Properly Sized Equipment During Normal Cooling Season

Albuquerque, NM	800-2200	Indianapolis, IN	600-1600
Atlantic City, NJ	500-800	Little Rock, AR	1400-2400
Birmingham, AL	1200-2200	Minneapolis, MN	400-800
Boston, MA	400-1200	New Orleans, LA	1400-2800
Burlington, VT	200-600	New York, NY	500-1000
Charlotte, NC	700-1100	Newark, NJ	400-900
Chicago, IL	500-1000	Oklahoma City, OK	1100-2000
Cleveland, OH	400-800	Pittsburgh, PA	900-1200
Cincinnati, OH	1000-1500	Rapid City, SD	800-1000
Columbia, SC	1200-1400	St. Joseph, MO	1000-1600
Corpus Christi, TX	2000-2500	St. Petersburg, FL	1500-2700
Dallas, TX	1200-1600	San Diego, CA	800-1700
Denver, CO	400-800	Savannah, GA	1200-1400
Des Moines, IA	600-1000	Seattle, WA	400-1200
Detroit, MI	700-1000	Syracuse, NY	200-1000
Duluth, MN	300-500	Trenton, NJ	800-1000
El Paso, TX	1000-1400	Tulsa, OK	1500-2200
Honolulu, HI	1500-3500	Washington, DC	700-1200

Cooling Load TD for Calculating Cooling Load from Sunlit Walls

North Latitude Wall Facing	Solar Time, hr 1	2	3	4	5	6	7	8	9	10	11	12	13	14	15	16	17	18	19	20	21	22	23	24	Hr of Maxi-mum CLTD	Mini-mum CLTD	Mini-mum CLTD	Differ-ence CLTD
Group A Walls																												
N	14	14	14	13	13	13	12	12	11	11	10	10	10	10	10	10	11	11	12	12	13	13	14	14	2	10	14	4
NE	19	19	19	18	17	17	16	15	15	15	15	15	16	16	17	18	18	18	19	19	20	20	20	20	22	15	20	5
E	24	24	23	23	22	21	20	19	19	18	19	19	20	21	22	23	24	24	25	25	25	25	25	25	22	18	25	7
SE	24	23	23	22	21	20	20	19	18	18	18	18	18	19	20	21	22	23	23	24	24	24	24	24	22	18	24	6
S	20	20	19	19	18	18	17	16	16	15	14	14	14	14	14	15	16	17	18	19	19	20	20	20	23	14	20	6
SW	25	25	25	24	24	23	22	21	20	19	19	18	17	17	17	17	18	19	20	22	23	24	25	25	24	17	25	8
W	27	27	26	26	25	24	24	23	22	21	20	19	19	18	18	18	18	19	20	22	23	25	26	26	1	18	27	9
NW	21	21	21	20	20	19	19	18	17	16	16	15	15	14	14	14	15	15	16	17	18	19	20	21	1	14	21	7
Group B Walls																												
N	15	14	14	13	12	11	11	10	9	9	9	8	9	9	9	10	11	12	13	14	14	15	15	15	24	8	15	7
NE	19	18	17	16	15	14	13	12	12	13	14	15	16	17	18	19	19	20	20	21	21	21	20	20	21	12	21	9
SE	23	22	21	20	18	17	16	15	14	14	15	16	18	20	21	23	24	25	26	26	26	26	25	24	21	14	26	12
S	21	20	19	18	17	15	14	13	12	11	11	11	11	12	14	15	17	19	20	21	22	22	22	21	23	11	22	11
SW	27	26	25	24	22	21	19	18	16	15	14	14	13	13	14	15	17	20	22	25	27	28	28	28	24	13	28	15
W	29	28	27	26	24	23	21	19	18	17	16	15	14	14	14	15	17	19	22	25	27	29	29	30	24	14	30	16
NW	23	22	21	20	19	18	17	15	14	13	12	12	12	11	12	12	13	15	17	19	21	22	23	23	24	11	23	9
Group C Walls																												
N	15	14	13	12	11	10	9	8	8	7	7	8	8	9	10	12	13	14	15	16	17	17	17	16	22	7	17	10
NE	19	17	16	14	13	11	10	10	11	13	15	17	19	20	21	22	22	23	23	23	23	22	21	20	20	10	23	13
E	22	21	19	17	15	14	12	12	14	16	19	22	25	27	29	29	30	30	30	29	28	27	26	24	18	12	30	18
SE	22	21	19	17	15	14	12	12	12	13	16	19	22	24	26	28	29	29	29	29	28	27	26	24	19	12	29	17
S	21	19	18	16	15	15	12	10	9	9	9	10	11	14	17	20	22	24	25	26	25	25	24	22	20	9	26	17
SW	29	27	25	22	20	18	16	15	13	12	11	11	11	13	15	18	22	26	29	32	33	33	32	31	22	11	33	22
W	31	29	27	25	22	20	18	16	14	13	12	12	12	13	14	16	20	24	29	32	35	35	35	33	22	12	35	23
NW	25	23	21	20	18	16	14	13	11	10	10	10	10	11	12	13	15	18	22	25	27	27	27	26	22	10	27	17
Group D Walls																												
N	15	13	12	10	9	7	6	6	6	6	6	7	8	10	12	13	15	17	18	19	19	19	18	16	21	6	19	13
NE	17	15	13	11	10	8	7	8	10	14	17	20	22	23	23	24	24	25	25	24	23	22	20	18	19	7	25	18
E	19	17	15	13	11	9	8	9	12	17	22	27	30	32	33	33	32	32	31	30	28	26	24	22	16	8	33	25
SE	20	17	15	13	11	10	8	8	10	13	17	22	26	29	31	32	32	32	31	30	28	26	24	22	17	8	32	24
S	19	17	15	13	11	9	8	7	6	6	7	9	12	16	20	24	27	29	29	29	27	26	24	22	19	6	29	23
SW	28	25	22	19	16	14	12	10	9	8	8	8	10	12	16	21	27	32	36	38	38	37	34	31	21	8	38	30
W	31	27	24	21	18	15	13	11	10	9	9	9	10	11	14	18	24	30	36	40	41	40	38	34	21	9	41	32

Cooling Load TD for Calculating Cooling Load from Sunlit Walls (Cont'd)

	Solar Time, hr																								Hr of Maximum CLTD	Minimum CLTD	Minimum CLTD	Difference CLTD
	1	2	3	4	5	6	7	8	9	10	11	12	13	14	15	16	17	18	19	20	21	22	23	24				
NW	25	22	19	17	14	12	10	9	8	7	7	8	9	10	12	14	18	22	27	31	32	32	30	27	22	7	32	25
Group E Walls																												
N	12	10	8	7	5	4	3	4	5	6	7	9	11	13	15	17	19	20	21	23	20	18	16	14	20	3	22	19
NE	13	11	9	7	6	4	5	9	15	20	24	25	25	26	26	26	26	26	25	24	22	19	17	15	16	4	26	22
E	14	12	10	8	6	5	6	11	18	26	33	36	38	37	36	34	33	32	30	28	25	22	20	17	13	5	38	33
SE	15	12	10	8	7	5	5	8	12	19	25	31	35	37	37	36	34	33	31	28	26	23	20	17	15	5	37	32
S	15	12	10	8	7	5	4	3	4	5	9	13	19	24	29	32	34	33	31	29	26	23	20	17	17	3	34	31
SW	22	18	15	12	10	8	6	5	5	6	7	9	12	18	24	32	38	43	45	44	40	35	30	26	19	5	45	40
W	25	21	17	14	11	9	7	6	6	6	7	9	11	14	20	27	36	43	49	49	45	40	34	29	20	6	49	43
NW	20	17	14	11	9	7	6	5	5	5	6	8	10	13	16	20	26	32	37	38	36	32	28	24	20	5	38	33

Cooling Load TD for Calculating Cooling Load from Flat Roofs

Roof No	Description of Construction	Weight lb/ft²	*U*-value Btu/(h·ft²·°F)	Solar Time, hr 1	2	3	4	5	6	7	8	9	10	11	12	13	14	15	16	17	18	19	20	21	22	23	24	Hour of Maximum CLTD	Minimum CLTD	Maximum CLTD	Difference CLTD
	Without Suspended Ceiling																														
1	Steel sheet with 1-in. (or 2-in.) insulation	7 (8)	0.213 (0.124)	1	−2	−3	−3	−5	−3	6	19	34	49	61	71	78	79	77	70	59	45	30	18	12	8	5	3	14	−5	79	84
2	1-in. wood with 1-in. insulation	8	0.170	6	3	0	−1	−3	−3	−2	4	14	27	39	52	62	70	74	74	70	62	51	38	28	20	14	9	16	−3	74	77
3	4-in. l.w concrete	18	0.213	9	5	2	0	−2	−3	−3	1	9	20	32	44	55	64	70	73	71	66	57	45	34	25	18	13	16	−3	73	76
4	2-in. h.w. concrete with 1-in. (or 2-in.) insulation	29	0.206 (0.122)	12	8	5	3	0	−1	−1	3	11	20	30	41	51	59	65	66	66	62	54	45	36	29	22	17	16	−1	67	68
5	1-in. wood with 2-in. insulation	19	0.109	3	0	−3	−4	−5	−7	−6	−3	5	16	27	39	49	57	63	64	62	57	48	37	26	18	11	7	16	−7	64	71
6	6-in. l.w. concrete	24	0.158	22	17	13	9	6	3	1	1	3	7	15	23	33	43	51	58	62	64	62	57	50	42	35	28	18	1	54	63
7	2.5-in. wood with 1-insulation	13	0.130	29	24	20	16	13	10	7	6	6	9	13	20	27	34	42	48	53	55	56	54	49	44	39	34	19	6	56	50
8	8-in. l.w. concrete	31	0.126	35	30	26	22	18	14	11	9	7	7	9	13	19	25	33	39	46	50	53	54	53	49	45	40	20	7	54	47
9	4-in. h.w. concrete with 1-in. (or 2-in.) insulation	52 (52)	0.200 (0.120)	25	22	18	15	12	9	8	8	10	14	20	26	33	40	46	50	53	53	52	48	43	38	34	30	18	8	53	45
10	2.5-in. wood with 2-in. insulation	13	0.093	30	26	23	19	16	13	10	9	8	9	13	17	23	29	36	41	46	49	51	50	47	43	39	35	19	8	51	43
11	Roof terrace system	75	0.106	34	31	28	25	22	19	16	14	13	13	15	18	22	26	31	36	40	44	45	46	45	43	40	37	20	13	46	33
12	6-in. h.w. concrete with 1-in. (or 2-in.) insulation	(75) 75	0.192 (0.117)	31	28	25	22	20	17	15	14	14	16	18	22	26	31	36	40	43	45	45	44	42	40	37	34	19	14	45	31
13	4-in. wood with 1-in. (or 2-in) insulation	17 (18)	0.106 (0.078)	38	36	33	30	28	25	22	20	18	17	16	17	18	21	24	28	32	36	39	41	43	43	42	40	22	16	43	27

Bin Method

The bin method of estimating yearly energy requirements is based on number of hours in five-degree increments. One could find, for instance, that an average winter's day had one hour in the zero bin, 4 hours in the 5 bin, 8 hours in the 10 bin, and 11 hours in the 15 bin. Seasonal requirements are determined by adding the total number of hours in each of the 5 bins.

Bin Method for Calculating Seasonal Heating Requirements

1 Mean D.B. Temp	2 No. Hours	3 Heat Loss Rate x 10^6	4 Internal Gain x 10^6	5 Setback Factor	6 Bin Heat Loss x 106
62	695	.465	.66	.313	---
57	633	.756	.66	.577	---
52	592	1.047	.66	.694	39.44
47	566	1.337	.66	.761	202.32
42	595	1.628	.66	.804	386.10
37	808	1.919	.66	.833	758.33
32	884	2.209	.65	.855	1086.17
27	618	2.500	.66	.872	939.36
22	377	2.791	.66	.885	682.38
17	248	3.082	.66	.896	521.17
12	131	3.372	.66	.905	313.31
7	61	3.663	.66	.913	163.74
2	17	3.954	.66	.919	50.55
−3	4	4.244	.66	.925	13.06
−8	1	4.535	.66	.929	3.55
				TOTAL	5,159.48

Assuming 80% efficiency, input energy = 6,450 x 10^6 Btu

Bin Method of Calculating Cooling Energy Requirements

1 D.B. Temp	2 No. Hours	3 W.B. Tamp	4 Sensible Gain x 10^6	5 Solar Gain x 10^6	6 Internal Gain x 10^6	7 W_o	8 Latent Gain x 10^6	9 Load x 10^6
92	33	75	31.8			.0148	11.36	
87	115	70	78.3			*		
82	261	68	103.7			*		
77	456	66	51.7			*		
72	669	62	−113.9			*		
67	791	59	−359.0			*		
62	690	56	−508.9			*		
TOTAL	3015		−716.3	934.7	1992.9		11.36	2,222.7

Hour-By-Hour Calculations

Still another, and more exact method, uses basically the same equations already employed to calculate the heat loss in our building. However, these equations are modified by weighting factors that reflect the thermal storage capacities of various construction materials and the characteristics of the energy distribution system. A computer program is used to perform these calculations, which employ hour-by-hour weather data.

Hourly Weather Occurrences

LOCATION	OUTDOOR TEMPERATURE, F																		
	72	67	62	57	52	47	42	37	32	27	22	17	12	7	2	-3	-8	-13	-18
Albany, NY	588	733	740	708	652	625	647	769	793	574	404	278	184	110	63	32	10	5	4
Albuquerque, NM	767	831	719	651	687	734	741	689	552	346	154	66	21	4	1	1			
Atlanta, GA	1185	926	823	784	735	676	598	468	271	112	44	19	8	2					
Bakersfield, CA	831	898	966	977	908	746	541	247	77	7									
Birmingham, AL	1138	908	805	742	668	614	528	433	292	143	69	17	6	3					
Bismark, ND	454	566	614	606	563	520	518	604	653	550	474	371	338	292	278	208	131	77	80
Boise, ID	492	575	643	702	786	798	878	829	522	307	148	53	26	14	6	2			
Boston, MA	676	819	804	781	766	757	828	848	674	429	256	151	74	35	4	9	1		
Buffalo, NY	646	772	760	700	666	624	647	756	849	602	426	267	170	815		24	2		
Burlington, VT	573	670	703	694	655	603	637	716	752	561	491	336	272	216	135	81	39	17	8
Casper, WY	423	532	592	642	606	670	782	831	806	683	495	325	200	116	73	45	30	15	5
Charleston, SC	1267	1090	889	787	651	576	434	321	192	79	27	5							
Charleston, WV	912	949	767	689	661	667	607	633	630	356	252	135	73	22	7	1			
Charlotte, NC	1115	908	839	752	730	684	634	515	360	166	64	23	5	2					
Chattanooga, TN	1021	895	775	722	713	679	642	553	414	228	113	45	4	4	2				
Chicago, IL	762	769	653	592	569	543	591	800	822	551	335	196	117	85	59	25	12	3	
Cincinnati, OH	879	843	726	639	611	599	627	698	711	460	249	131	68	44	18	8	2		
Cleveland, OH	763	831	723	641	638	607	620	754	806	578	355	201	111	47	22	11	2		
Columbus, OH	774	820	720	648	622	603	658	730	772	502	280	169	94	40	20	10	4	1	
Corpus Christi, TX	1175	1041	748	551	444	302	180	83	27	9	3								
Dallas, TX	831	795	693	656	629	576	504	371	231	91	34	17	4	1					
Denver, CO	549	684	783	731	678	704	692	717	721	553	359	216	119	78	36	22	6	1	1
Des Moines, IA	707	751	681	600	585	512	510	627	747	557	405	281	211	152	104	59	23	8	1
Detroit, MI	721	783	695	633	592	566	595	808	884	618	377	248	131	61	17	4	1		
El Paso, TX	933	839	749	760	687	611	494	369	233	104	34	10	2						
Ft. Wayne, IN	728	777	699	608	569	552	601	725	905	596	381	205	124	69	40	19	6	1	
Fresno, CA	709	803	921	1006	1036	952	673	426	168	34									
Grand Rapids, MI	634	739	712	647	571	565	554	742	938	690	469	293	172	78	31	10	1	1	
Great Falls, MT	407	520	636	754	822	830	832	813	698	533	355	218	167	136	118	101	68	51	62
Harrisburg, PA	807	824	737	692	635	659	722	888	749	427	222	125	52	18	4	1			
Hartford, CT	617	755	751	752	649	575	683	807	825	552	370	233	153	77	33	11	3	2	
Houston, TX	1172	980	772	681	570	452	291	141	64	18	4	2							
Indianapolis, IN	821	815	722	585	586	579	605	712	791	551	293	152	97	60	35	13	3	2	
Jackson, MS	1168	922	790	677	618	605	484	367	224	103	41	6	2	2	1				
Jacksonville, FL	1334	975	879	692	530	355	288	154	83	24	2								

Hourly Weather Occurrences (Cont'd)

Kansas City, MO	761	723	601	572	553	562	628	625	591	407	265	175	99	51	21	4			
Knoxville, TN	1056	889	746	675	672	689	648	590	456	217	101	41	21	7	2				
Las Vegas, NV	651	644	699	786	769	716	591	396	194	44	7	1							
Little Rock, AR	940	803	725	672	638	669	605	509	363	172	50	25	5	1					
Los Angeles, CA	881	1654	2193	1904	1054	428	107	10											
Louisville, KY	869	758	693	654	619	634	649	703	631	332	169	97	45	25	8	3	1		
Lubbock, TX	833	829	688	700	642	618	620	546	490	346	180	86	33	7	5	1			
Memphis, TN	977	798	715	690	618	633	614	532	374	196	74	25	10	4					
Miami, FL	1705	810	452	277	147	71	26	4											
Milwaukee, WI	597	753	749	634	585	591	611	774	913	659	421	285	176	116	83	47	18	4	3
Minneapolis, MN	621	690	695	602	588	482	500	560	632	609	514	383	311	246	186	119	62	31	16
Mobile, AL	1411	1038	882	698	609	506	377	214	109	49	7	3							
Nashville, TN	933	838	738	697	637	619	627	565	463	263	132	67	28	9	3	1	1		
New Orleans, LA	1189	987	850	692	621	449	282	128	47	9	2								
New York, NY	926	877	754	745	722	796	838	858	603	330	188	2	26	10	1				
Oklahoma City, OK	881	769	717	643	645	611	641	570	468	287	173	77	36	12	3	1			
Omaha, NB	726	721	606	558	539	543	543	655	663	511	390	287	189	135	93	40	15	1	
Philadelphia, PA	863	809	735	710	663	701	758	818	654	335	189	100	32	9					
Phoenix, AZ	762	776	767	769	659	540	391	182	57	8									
Pittsburgh, PA	722	910	799	678	637	587	631	688	569	774	360	233	159	60	30	7	1		
Portland, ME	407	627	780	808	760	748	722	839	820	599	408	293	190	109	60	29	15	5	1
Portland, OR	373	581	1001	1316	1274	1271	1238	772	343	123	40	10	4	1					
Raleigh, NC	1087	937	848	762	707	672	638	527	410	236	103	38	11	1					
Reno, NV	418	477	572	690	845	909	890	829	733	530	387	277	101	37	15	4	1		
Richmond, VA	953	850	784	745	690	673	699	632	478	285	138	67	19	2	1				
Sacramento, CA	630	773	1071	1329	1298	1049	701	355	93	8									
Salt Lake City, UT	569	615	614	635	682	685	755	831	798	564	328	158	80	41	16	2			
San Antonio, TX	1086	943	789	669	569	445	387	190	94	31	11	4	1	1					
San Francisco, CA	285	665	1264	2341	2341	1153	449	99	10										
Seattle, WA	258	448	750	1272	1462	1445	1408	914	427	104	39	20	3						
Shreveport, LA	1063	886	772	679	619	609	516	361	200	72	23	6	2						
Sioux Falls, SD	566	684	669	605	522	498	501	625	712	585	520	448	293	208	152	102	59	43	18
St. Louis, MO	823	728	646	575	585	578	620	671	650	411	219	134	77	40	15	7	1		
Syracuse, NY	627	735	723	717	656	641	651	720	830	547	392	282	190	102	55	23	5	2	2
Tampa, FL	1387	1187	877	570	345	216	137	48	10	1									
Waco, TX	909	830	701	622	651	558	501	354	216	84	24	3	1						
Washington, DC	960	766	740	673	690	684	790	744	542	254	138	54	17	2					
Wichita, KS	758	709	641	603	589	592	611	584	607	426	273	161	85	45	14	3	1		

Chapter 9
TESTING AND BALANCING VARIABLE AIR VOLUME SYSTEMS

BALANCING VAV SYSTEMS

When the thermostats are set to maximum flow in a VAV system it becomes, in effect, a constant air volume system and can be balanced as such.

The general concept of VAV balancing involves:

- VAV systems are balanced on maximum air flow and then spot checked on minimum flow.
- VAV terminals must be checked to make sure they are calibrated, have sufficient pressure at the intakes and are maintaining correct flows at maximum and minimum.
- Supply outlets on the discharge side of the VAV terminals must be balanced for the correct volume of air.
- The static pressure sensor or air flow monitor in the main duct, which is sensing the variations in duct pressure, must be checked on maximum and minimum flows and correlated with the fan volume control.
- Supply and return air fan volumes and coordinated tracking must be checked on maximum and minimum flows.
- Ductwork must be sealed in pressurized systems and be able to hold minimum pressures at terminal intakes. Ductwork air leaks must be defeated and sealed.
- The outside air and return air dampers must be checked on maximum and minimum flows.

BASIC PRINCIPLES OF BALANCING AIR SYSTEMS

1. Start at heart of the system, which is at the fan, and make sure it is pumping correctly before balancing duct runs and outlets.
2. Make sure the system is clean before balancing, that all filters, coils, strainers, duct runs, dampers, louvers, etc. are clean and unclogged.
3. Make sure the system is open before balancing, that all dampers behind grilles and supply diffusers, manual balancing dampers and fire dampers are all open and that the control air dampers are set in the correct positions.
4. Make sure the air distribution system is properly sealed and that no duct end caps are left off, no duct runs are unfinished and no outlets are not connected. Make sure that connections and seams are sealed, if required.
5. Balance on maximum air flow mode and with the maximum resistance load imposed such as with wet cooling coils or assimilated conditions.
6. Proportionate balance low pressure side of VAV terminals. It insures that the least amount of resistance is being input to achieve the required air balance, it requires the least amount of balancing time and maintains the proportionate amounts of air at maximum and minimum flow and at all points in between.

OVERALL VAV BALANCING PROCEDURE

1. Prepare test reports listing each terminal with its outlets. Add up the outlet CFM's and check against each box. Add up the total terminal CFM's and check against the fan CFM for diversity. If there is diversity, plan what segments will be alternately shut down and in what order.
2. Check that the building and system are complete and functionally ready for balancing.
3. Go through the standard checkout of equipment, starter, fan, fan drives, bearings, filters, coils, etc.
4. Bump fan for rotation.
5. Check out the outside and return air damper control dampers for proper hookup and operation.

6. Close inlet van dampers on centrifugal fans, or feather blades down on vaneaxial fan, on new systems. If the motor is controlled by a frequency inverter, set it to 50 percent level. Slowly open inlet damper or vaneaxial blades or inverter on startup noting possible bursting or collapsing of duct work.
7. Set all stats to maximum flow which is usually maximum cooling, 55°F.
8. Make sure ductwork is sealed and has been leak tested.
9. Spot check minimum and maximum settings on controllers on terminals.
10. Make sure all supply and return outlets and duct dampers are open.

Balancing VAV Systems

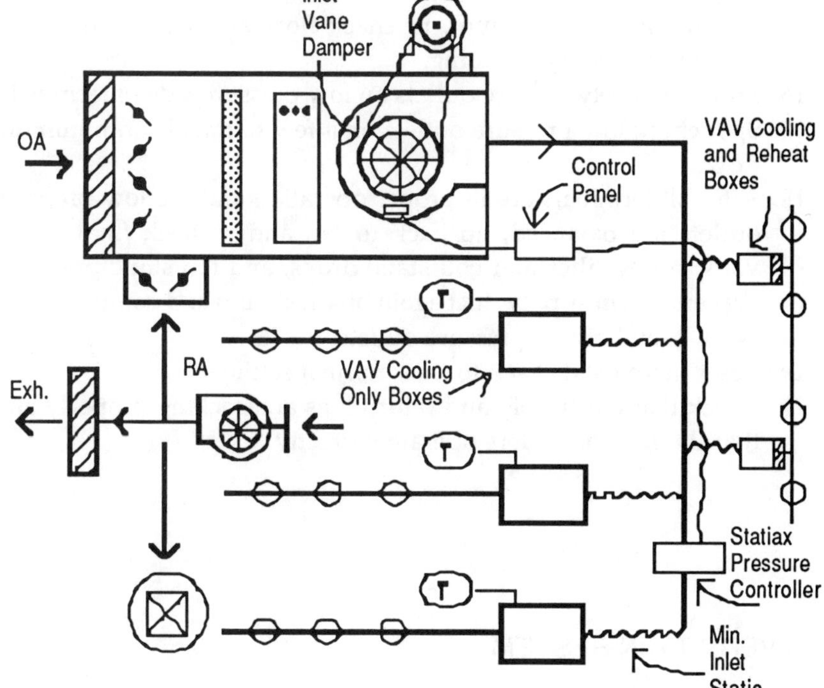

Figure 9-1
A typical VAV system contains some type of fan air volume controller, central equipment, appropriate interior and exterior terminals, duct pressure sensor, control panel and space thermostats controlling VAV terminals.

11. Turn on the fans, the return air first and then the supply. Slowly open vortex damper, vaneaxial blades or inverter to 100 percent flow.
12. Take startup readings: amps, volts and fan rpm. Read the fan suction and discharge pressures to make sure they are sufficient to operate the system. Check static pressure drops across filters and coils.
13. Spot check inlets of end terminals furthest away from fan to make sure there is enough pressure to operate them on pneumatic systems.
14. If there is enough pressure to operate all the terminals and everything is in proper operation, read the total air flow at the discharge of the supply fan.
15. Read air flows out of terminals on the low pressure side on maximum and minimum, to check flow against terminal unit controller setting and rating.
16. Proportionately balance outlets on low pressure side of terminals.
17. Spot check low pressure outlets on a few terminals on minimum flow.
18. After all terminals are in proper operation and the low pressure outlets are balanced, go back to fan and recheck final amps, volts, rpm's, filter and coil static drops, and fan static pressure. Traverse main supply duct again on maximum and compare CFM against total of low pressure outlets.
19. Reset thermostats back to their normal settings.
20. Check that the supply and return fans are tracking properly and that the fan controllers operate correctly.

DIVERSITY IN SYSTEM

If the total fan CFM is less than the total of all the CFM's of the terminals, there is diversity in the system and the balancing procedure must be handled differently than when the CFM's of the fan and terminals are the same.

Diversities between fan and terminal CFM's generally are in the magnitude of 20 to 30 percent; that is, the fan might only have the capacity to deliver 30,000 CFM while the total of the terminals is 40,000 CFM.

Diversity simply means that a 100 percent peak load will never exit in all areas served by the system simultaneously. The east zone peaks first, while the south and west are below their peaks. Then the east tapers off and the south peaks. Later in the day the south reduces and the west peaks.

Interior zones might peak mid-mornings and mid-afternoons. Therefore the total CFM of all the terminals is never all needed at one time and the fan CFM need not really be any more than the maximum simultaneous load.

The procedure for balancing a VAV system with diversity is as follows:

1. Since the fan cannot handle all the terminals at one time, different segments, equal or greater than the difference between the fan and terminal CFM's, must be shut down one at a time, and the rest of the system, which is roughly equal to the total fan CFM, is checked out.

2. Generally, following the sun and shutting down one major segment at a time, the east first, then the south and west and finally the interior segments in sequence while the others remain open, works well.

This procedure assures that there is sufficient pressure at the terminals to operate them with some other segment shut down, but it does not absolutely guarantee correct operation for normal operation where various mixtures of boxes might be shut down or open between the segments. Actual operation of the building under different conditions must be checked also.

The proportionate balance of the outlets on the low pressure side of the terminals will hold no matter which other combination of boxes shut down so long as there is sufficient intake pressure if pressure independent terminals are used.

NO DIVERSITY IN SYSTEM

If the total fan and terminal CFM's are the same, the procedure for balancing the VAV system is the same as balancing a constant volume high pressure system with the noted exceptions.

PRELIMINARIES

Study the plans, specifications and equipment drawings to become familiar with the systems. A determination must be made of the best method to balance the systems, and appropriate instruments must be selected and checked out.

Prepare Test Reports, Study Plans, and Specifications. The first stage in the testing and balancing procedure is the preparation of test reports. Equipment test report sheets must be completed for each system. Outlets must be listed on air balance sheets in the sequence of balancing together with their types, sizes, Ak factors if required, design air quantities and velocities.

Check That Building and Systems are Complete and Operational. After the reports are prepared, inspect the job site to see that the building and systems are architecturally, mechanically and electrically ready to be balanced and that they are complete and functional.

Invariably, new buildings may only be half ready when balancing starts, and in fact, it is the balancer's quality control check that uncovers a multitude of missing or incorrect items. As the balancing technician inspects each system he must report the inadequacies, see that corrective action is taken and move on to the systems that are ready for balancing.

After you have determined which systems are truly complete choose the first one to balance and proceed with an in-depth equipment check out.

CHECK HEART OF SYSTEM FIRST

Start the actual testing and balancing process at the heart of HVAC system, **the fan.**

If the heart isn't working right, the rest of the body can't perform as it should. Just as a doctor checks your blood pressure and pulse rate, you must check the fan's pressure and RPM rate.

The motor on the fan is the organ that drives the fan and its electrical characteristics must be checked out and it must be protected.

Hence, the first phase in the testing and balancing process is to check five items at the fan:

1. Motor amp draw thermal overloads
2. Fan rpm
3. Fan suction and discharge pressures
4. Pressure drops across components
5. Total air flow at fan

Then, after the heart of the system is checked, adjusted and is running properly, **and only then**, should the outlets and duct runs be read with instruments and balanced.

CHECK MOTOR AND STARTER

1. Motor Name Plate. Since the weakest link in the system is the motor, it is imperative that it be protected. Check the motor name plate first for maximum amp load, voltage, phase, rpm, service factor and other data. Record and compare with the design requirements written on the equipment sheets. If there are discrepancies in the voltage, phase or rpm they must be reconciled.
2. Thermal Overloads. Go to the starter next and check that the thermal overloads are installed and that they are the correct size. In a 3-phase system there must be three overloads, one for each line. If they are not installed, do not test the system until they are!

   The thermal overloads must be the correct size and not exceed the motor name plate amps. For example if the maximum name plate amps are 12.0, the thermal overload must be rated for a maximum of 12 amps, plus or minus a few tenths. The correct size overload is normally on a chart on the inside of the starter cover.

FAN TEST REPORT

Job North High School Job No C-150 Date Aug. 1, 1990
Location ______ System S-2
Equipment Location Mezzanine Serves Lunchroom Tested By: HW

☑ Air Handling Unit ☐ Roof Top Unit ☐ Furnace ☐ Supply Fan ☐ Exhaust Fan ☐ Pkg Unit
☑ LP ☐ MP ☐ HP ☑ Constant Volume ☐ VAV

FAN DATA	
Manufacturer	Barry
Model Size	AF 7245 DWDI
Type Fan	☑ Centrigal ☐ Roof Exhaust ☐ Inline ☐ Vane Axial ☐ Prop.
Type Wheel	☐ Backward Incline ☑ Air Foil ☐ Forward Curve ☐ Paddle Wheel
Wheel:	☑ Alignment OK ☑ Gap ☑ Fastened ☑ Clean
Belts (2) B131	C to C Distance 52"
Pulleys: Fan Dia. 10"	Mot. Dia. 5"
Motor Movement 2" ±	Belts
Bearings ✓	☑ Cut Off Plate OK

MOTOR		
Manufacturer GE	Serial No.	
Frame No. 184T	Type Frame ☐ T ☑ U	
Service Factor: 1.15	Rated	Actual
HP, Nameplate	5	5
BHP $[HPnp \times \frac{Aa}{Ar} \times \frac{Va}{Vr}]$	3.53	3.84
Amps, L_1 L_2 L_3	15.2	12.2
Voltage, L_1 L_2 L_3	230V	220V
RPM	1750	1750
Phase	3	3

FAN PERFORMANCE		
	Design	Actual
Fan CFM	9,800	10,160/104%
Outlet CFM Total	9,800	9,756/99%
Fan RRM	985	992
Fan S.P.	1½"	1.6"

STARTER	
Manufacturer GE	Model 141R
Starter Size 0	Class 1
Overload: Required Size	CR 15.4
Actual	CR 15.4

CONDITIONS			
Vortex Damper Position	—		
Outside Air Damper Setting	4000 CFM		
Return Air Damper Setting	6160 CFM		
Filter Conditions	Clean		
Coil Conditions	Clean		
Temperatures			
OA: @ 40% 40 F	DB	WB	RH
RA: 70 F	DB	WB	RH
Mixed Air:	DB	WB	RH
Discharge	DB	WB	RH
Space:	DB	WB	RH
Duct Temp. Drop	DB		

STATIC PRESSURE DROPS			
	Upstream	Downstream	Total Drop
Filter	.2"	.4"	.2"
Heat. Coil	.4	.6	.2
Cool. Coil	.6	1.0	.4
Fan Inlet			1.0
Fan Discharge			.6
Total Fan S.P.			1.6"

Remarks ______

PROBLEMS:
☐ Too much air ☐ SP Low ☐ Too Hot
☐ Too little air ☐ SP High ☐ Too Cold
☐ Air Noises ☐ Fan Noises
☐ Oversized equipment ☐ Undersized Equipment
☐ Other ______

Figure 9-2

OUTLET TEST REPORT

OUTLET AIR BALANCE REPORT

Project North High School Job No. C-150 Date Aug. 1, 1990

Location Classroom Building System S-1, VAV, MP

Instruments Used Velometer, Magnehelic Gauge Tested by: HW

ROOM AREA SERVED	OPENING				REQUIRED		PRELIMINARY			FINAL	
	No.	Model	Size	A_k	Vel	CFM				Vel	CFM
Rm 101, Box	A-1				—	600	Pitot Traverse				(605)
Outlets		PS	12" ϕ	.66	455	300					305
		PS	12"	.66	455	300					290
Rm 102, Box	A-2				—	600					
		PS	12"	.66	455	300					310
		PS	12"	.66	455	300					300
Rm 103, Box	A-3					600					
		PS	12"	.66	455	300					280
		PS	12"	.66	455	300					305
Rm 104, Box	A-4					600					
		PS	12"	.66	455	300					305
		PS	12"	.66	455	300					310
Rm 105, Box	A-5					600					
		PS	12"	.66	455	300					285
		PS	12"	.66	455	300					305
Rm 106, Box	A-6					600					
		PS	12"	.66	455	300					290
		PS	12"	.66	455	300					300
Rm 107, Box	A-7					600					
		PS	12"	.66	455	300					310
		PS	12"	.66	455	300					305
(continued p. 3)											

Remarks Minimums on boxes: SP .4"
CFM 25%

Figure 9-3

Locate the maximum amps in the column and read the size heater required next to it. Usually the heater number is stamped on the face of the heater itself and is visible when installed.

Inspect Fan Components

1. Fan Wheel. Inspect the fan wheel next. Is it the correct type and size? On centrifugals it could be one of four basic types, backward inclined, air foil, forward curve or paddle wheel.
   Is the fan wheel installed correctly? Sometimes the factory installs a fan wheel backwards in a fan or if the fan is knocked down and assembled on the job site it frequently will be installed backwards.
   Is the gap and center line alignment between the wheel and the inlet cone on centrifugal fans correct? This can cause internal fan cycling and major havoc on the fan performance reducing air flow 30, 40 or 50%. See fan section on centrifugal fan wheel gaps and alignment.
   Check to see that the wheel is securely fastened to the shaft. Check that the bearings are greased properly if they are not the permanently lubricated type.

2. Drives. Inspect the drives. Is the belt tension correct? On multi-belted drives is the tension the same on each belt? If not it could indicate that the belts are of different lengths and are not a matched set.
   Is the alignment correct? Cockeyed belts wear out super fast and do not efficiently transmit horsepower.
   Make a rough metal calculation of the pulley diameter ratio and compare with the motor fan rpm ratio. Catastrophes have occurred when new or remodeled systems were first turned on. Ducts have burst apart or collapsed faster than you could whip your hand back to the starter.
   This rough metal check involves dividing the fan pulley diameter by the motor pulley diameter and comparing that ratio to the motor rpm divided by the fan rpm.
   For example, if fan pulley is 10 inches in diameter and the motor 5, divide 10 by 5 and the ratio is 2. Now compare this with motor

rpm, 1800 divided by a design fan rpm of 900. This is also a ratio of 2 and all is well. If the ratios do not match within 5 or 10 percent the discrepancy must be reconciled before turning the fan on.

Fan Pulley Dia. = Design Motor rpm
Motor Pulley dia. = Design Fan rpm

Example: $\frac{10\text{" Dia.}}{5\text{" Dia.}} = \frac{1800\text{ rpm}}{900\text{ rpm}} = 2$

Record pulley diameters, belt sizes, the true center distance from the motor shaft to the fan shaft and available motor movement back and forth for tension adjustment in case drives must be changed.

3. Bump the fan to check the rotation of the wheel. Frequently motors are wired in reverse. To reverse the direction of a three-phase motor, switch two leads at the motor or starter. For single-phase starters check the motor wiring diagram. Bumping the fan simply means turn the fan on and off again quickly.

INSPECT SYSTEM COMPONENTS

1. Filters. Inspect the filters to see that they are installed and clean. On new jobs, if they are a temporary construction set, replace with the permanent set. If a permanent set, make sure they are not excessively dirty or clogged.

2. Cooling and Heating Coils. Check the cooling and heating coils. In built-up housings, are they properly blanked off all around the tops and bottoms and sides so air does by bypass the coil? Are there large gaps where the piping connections protrude through the side of the housing? If so, seal properly.
   Check if the coils are clean. If the system must be balanced with the heating or cooling on, are the coils and control valves in proper operation. If balancing must be done in a cooling mode and the cooling system is not operable for whatever reason, portions

of the coil face area can be blocked off with cardboard or polyethylene.

3. Automatic Dampers. The next step in the system component checkout is to check and set the automatic dampers in their balancing positions. There are two approaches in settings of outside air, return air and exhaust control dampers.

   The first approach, with a separate RA fan, is to set the outside air to 100 percent open, the return air to 100 percent closed, and the exhaust dampers to 100 percent. This approach is based on assuring that the supply fan can handle the full volume of air without the help of the return air fan in this extreme condition. After balancing on 100 percent the OA damper is swung to its minimum position and outlets are spot checked.

   The second approach is just the reverse of the above. The OA is set at its minimum position and the return air in its maximum, then after balancing this way the system is spot checked in the 100 percent OA mode.

   If there is no separate return air fan where the supply fan is handling both the supply and return, the maximum load on the fan is when the OA is at minimum and return air is at maximum. This is the mode that the balancing should be done in. Spot checking must be done in the maximum OA and minimum RA positions.

   If there are automatically controlled face and bypass dampers by a heating coil, the face damper should be 100 percent open and the bypass closed.

   If there are automatically controlled vortex dampers on the intake of a centrifugal fan, as with medium and high pressure VAV systems, they should be closed completely and then upon startup opened slowly, to prevent possible bursting of ductwork.

4. Outlet and Ductwork Dampers. After the central equipment is set up, go through the spaces served by the system and shine a flashlight through all outlets to make sure that all the grille and ceiling diffuser dampers are 100 percent open before turning the system on.

Check that splitter dampers are positioned at roughly a 30- to 45-degree angle and that other manual volume dampers and fire dampers are 100 percent open.

5. Thermostat Settings. Set stat to maximum cooling, usually 55 degrees, so that the coil is wetted and the system is balanced under its maximum load.

6. VAV Terminals. Check that the VAV terminals are set properly for the correct maximum and minimum flows.

TAKE FAN READINGS

1. Start Up. After completing the inspection and set up of the equipment and ductwork, turn on the equipment to be balanced plus all other systems that serve the same area and take start up readings.
   Upon start up listen for bursting or collapsing ducts, a rubbing fan wheel, motor or bearing noises, or rumbling or clanging of any type. Observe the operation of the automatic dampers. If something erratic is seen or heard, turn the fan off immediately, check out and rectify the problem before proceeding.

2. Amp and Volt Reading. Since a motor can be burned up so quickly, the first thing to do after starting the equipment is to check the amp draw, to make sure it is not exceeding maximum motor amps, and check the voltage to confirm it is in the correct range.
   This is normally done with a volt-ammeter at the starter. The jaws are clamped around each wire, one at a time, and the amps read. Then the probes are used to read voltage across terminals.
   If there is a big difference between the amps on the legs, or if the voltage deviates greatly from design, or fluctuates, there may be electrical system problems which have to be resolved before you can proceed any further in testing the system. (See Figure 9-4.)

3. Rpm Reading. Immediately after the amp-volt reading, check the fan rpm to see that it is approximately as per the design. Use a tachometer as described in Chapter 4.

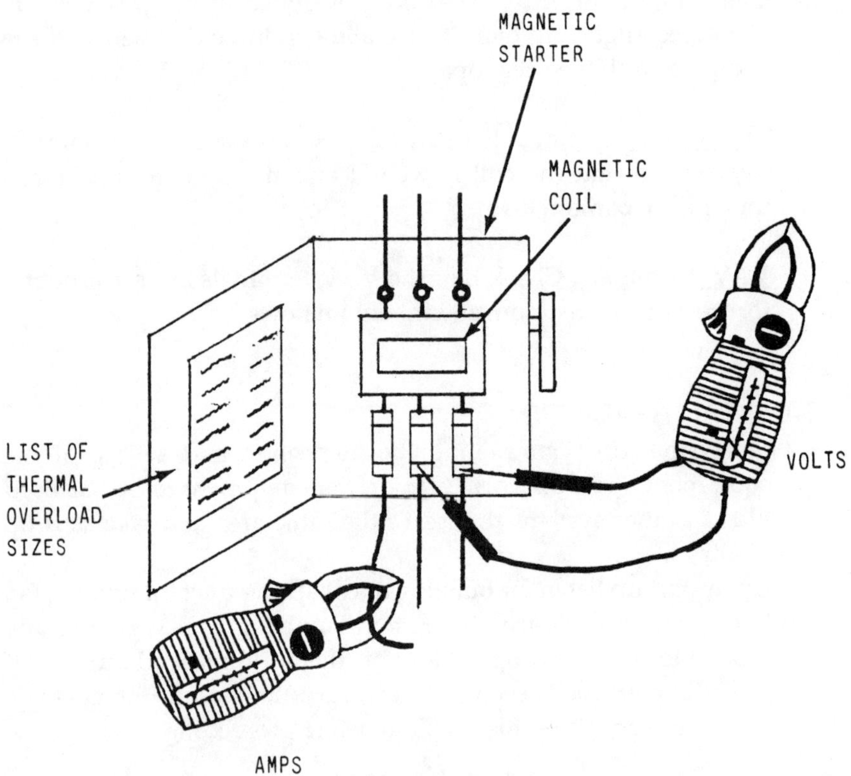

Figure 9-4.

If the rpm of the fan is grossly higher or lower than design, check the following:

a) Check the motor rpm to see if a wrong speed motor was installed.
b) Check the pulley diameters to see if you have the correct diameter ratio.
c) Check if the blueprints, fan drawings or test report sheets are in error, or if there was a change.

The drive belts may also simply be riding too high or low in variable pitch motor pulley. If the amp draw permits it, change the variable pitch sheave to get the fan at the correct rpm.

4. Fan Suction and Discharge Pressures. Read the fan suction and discharge static pressures next and add them together for the total fan static pressure. For example, a typical suction pressure may be 1 inch and the discharge .5 inches. This wold be a total of 1.5 inches.

5. Pressure Drops Across Suction Side Components. The pressure drops across the filters and coils should be taken next. They serve as a check against the design engineer's calculations and equipment manufacturer's catalog ratings, for possible flow problems and analysis and for future reference.

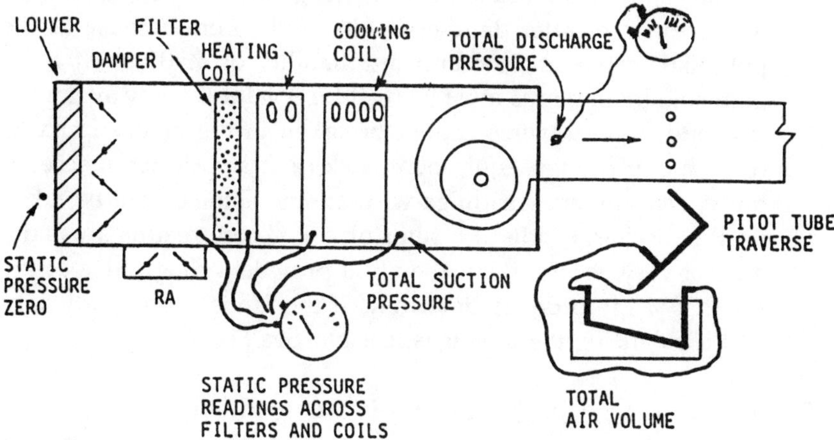

Figure 9-5.

The pressure drops across the filters, coils and control dam can be taken with a magnehelic gauge with a 0- to 1-inch or 0- to 2-inch scale. Drill holes at the intake of the filter between the filter and first coil, between the coils and at the fan intake. Take individual static pressure readings at each point and subtract upstream from downstream readings for arriving at the drops.

6. Total Air Flow. Knowing three characteristics of the fan performance out of the five, the amp draw and rpm, a third critical aspect is checked at this point.

Check the total air flow from the supply fan to see if you have approximately the correct amount of air to start off with before balancing the outlets.

The main exception to this procedure is small systems with few outlets where it is easier and faster to read all the outlets and total them up, than to take a total air flow reading at the fan. Pitot shots also depend on accessibility of main ducts, complications of routing and number of floors or different spaces the system serves.

The most accurate method of taking a total air flow reading is with a pitot tube traverse in a straight run of ductwork five to ten times the width of the duct. Readings must not be taken in or near fittings or after dampers, coils, and so on, because of the potential turbulent flow and unreliability of readings at these points. Enough points must be read for a valid velocity average.

If a pitot tube traverse cannot be taken in the main discharge duct due to fittings, equipment, lack of straight duct, inaccessibility, etc., traverse readings with an anemometer can be taken on the discharge side of a filter or coil. These readings usually are not very accurate, but they will provide a rough idea of the total CFM's in order to determine if the fan is running all right and if balancing the outlets is feasible. See Fig. 9-6.

7. Stratification Check. Air stratification through coils, filters, louvers, dampers, etc., can cause coil freeze up, under or over heating or cooling, and great energy inefficiency. If the arrangement of the outside air flow and return air flow into the mixing plenum gives any indication that there might be poor mixing of the air, resulting in temperature or velocity stratification, check out for stratification.

CHECK VAV TERMINALS

Check that there is sufficient pressure at the intake of VAV terminals to operate the terminals on pneumatic systems. Read outlets with a flow hood and add up the CFM's to make sure the total maximum flow is correct. See Fig. 9-7.

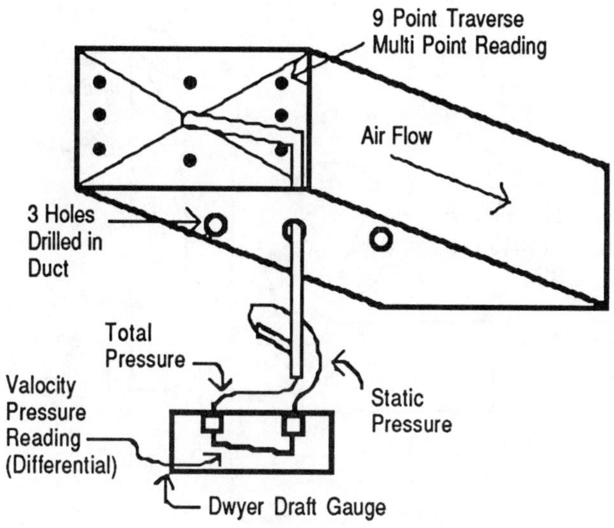

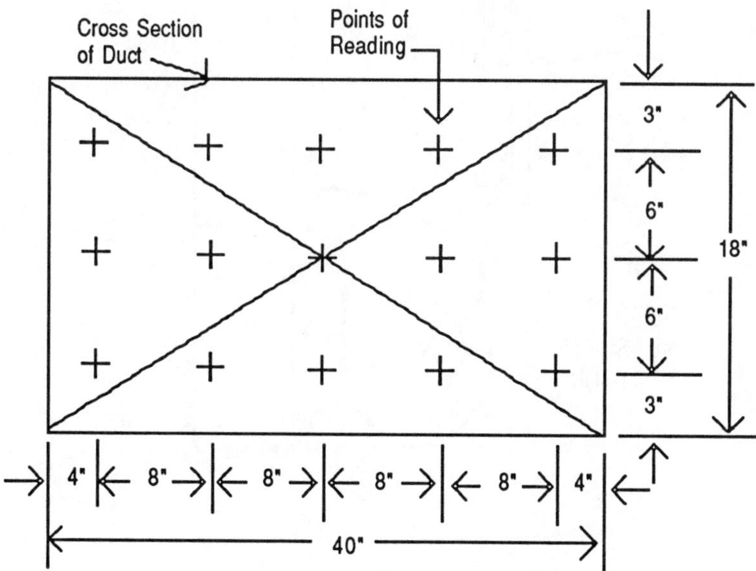

Example layout of points of reading for pitot tube traverse.

Figure 9-6.

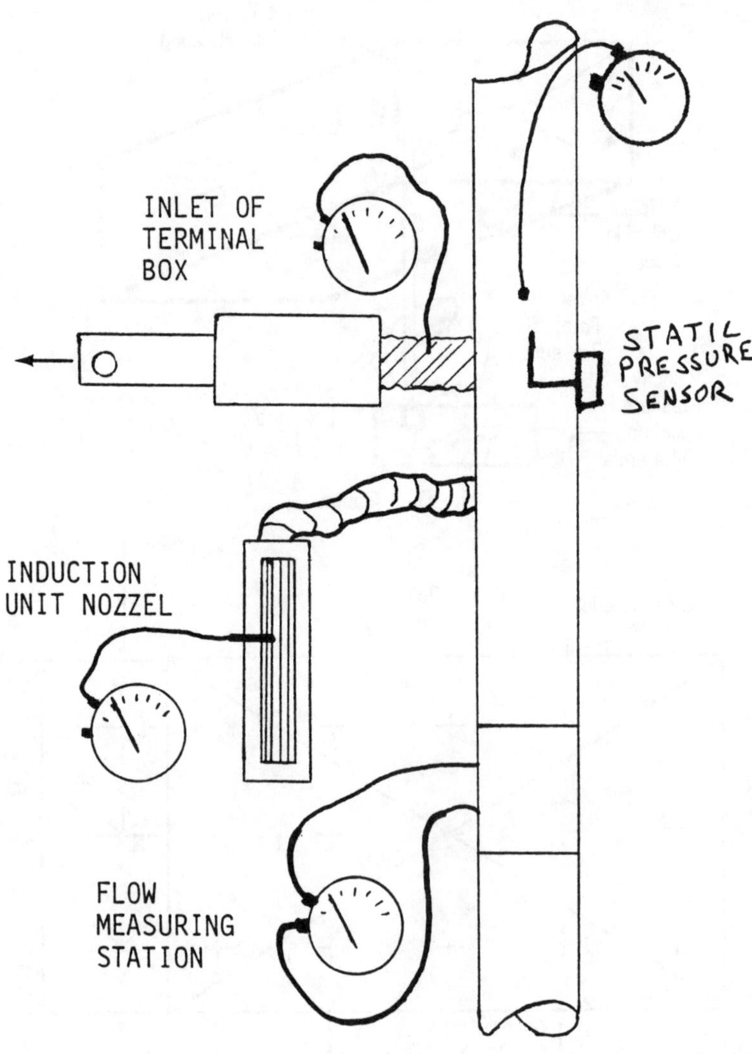
INLET OF
TERMINAL
BOX
STATIC
PRESSURE
SENSOR
INDUCTION
UNIT NOZZEL
FLOW
MEASURING
STATION

Figure 9-7.

PROPORTIONATE BALANCE OUTLETS AND DUCT RUNS

After the equipment is found to be correct and the total CFM in the right range, the outlets and branch ducts can be balanced.

First walk through the various areas served by the system to see if there are any problems with temperatures, drafts, air noises, etc. Spot check some end, middle and starting outlets to roughly determine the extent of imbalance. Then proceed with the balancing.

The most effective method of balancing is the "proportionate" method. This method results in the least amount of energy usage by the fan. In proportionate balancing all outlets and branch ducts in the system, starting with those furthest from the fan, are brought to about the same percent of design, give or take 5 percent. The method of doing this will be covered in the following section.

Constant volume, single zone, low pressure supply systems and exhaust systems can be proportionately balanced. High pressure systems involve a slightly different sequence of work for the high pressure side of the system, as will be described in the procedures for balancing the different systems.

FINAL SETTINGS AND READINGS AT FAN

After balancing the outlets, VAV terminals and low pressure systems are balanced at maximum flow. Return to the fan and recheck the total CFM amps and discharge static pressure.

If the total of the outlet CFM's is ten or more percent or higher or lower than design, the fan rpm should be increased or appropriately decreased to get as close to 100% of design as possible.

If the outlets were correctly balanced, their flows will all increase or decrease roughly the same percentage as the fan CFM. For example, if the fan is increased 12 percent, the outlets will each do likewise. Or if the fan flow is decreased 15 percent, the outlets will also.

If the CFM must be increased, check the actual amps against the motor full-rated amps to make sure they are not exceeded. Check to see if there is any room on the variable pitch drive on the motor to alter the rpm.

Using **Fan Law Number 1**, calculate the new rpm needed to achieve the new CFM. Calculate also what the new static pressure and break horsepower are. Compare BHP with actual HP of motor. Determine if you need new belts or sheaves. Check motor movement forward and backward in regard to whether belts can be reused or not. If a new motor is indicated consider if you can live with less CFM to retain existing motor.

After changing the CFM at the fan, spot check outlets to verify they have increased or decreased proportionately.

Unsealed low pressure ductwork might leak anywhere from 5 to 15 percent of the air flow. The average is about 8 percent. The fan CFM, under these conditions of leakage, will not match the total flow at the outlets and will generally run about 8 percent.

Medium and high pressure ductwork which has been properly sealed and leak tested should not leak more than 1 percent. The fan and total outlet CFM should be relatively the same in these situations.

TESTING AND BALANCING INSTRUMENTS

The first few sets of instruments we will look at are the one required for checking out the fan, electrical, rpm and pressure reading.

ELECTRICAL READING INSTRUMENTS

The clamp-on volt ammeter is used in balancing to read the actual amps and volts of fan and pump motors.

These readings are checked to insure motor protection and then determine the actual fan BHP. This is the first of five fan performance characteristics needed to evaluate fan performance or to change the fan flow. The readings are generally taken at the starter, but they can also be taken at disconnect switches and motors.

There are two basic types of clamp-on meters used today by balancers. One is the older rotary scale type and the other is the digital. Digital types may incorporate the ability to ready the starting surge current as well as normal running currents. Some meters may also incorporate fused overload protection and the ability to read ohms. See Figure. 9-1.

The clamp-on volt-ammeter reads amps by measuring the magnetic field around the conductor as current flows through it. The jaws of the ammeter form a continuous core around the conductor like a transformer and the magnetic field around the wire creates a secondary current in the meter which is proportioned to the scales on the meter.

Amp Readings

Amperage on a single-phase or three-phase motor is determined by clamping the jaws around each wire one at a time. Generally the best place to take readings is at the starter. With the starter cover off, pull the wires out of the starter box enough to wrap the jaws around them.

Voltage Readings

Voltage readings are also normally taken at the starter. This is done by inserting the voltage leads into jacks at the bottom of meter, touching one probe to one of the terminals in the starter, touching the other probe to a second terminal and then reading the voltage.

On a three-phase starter, three readings are taken, terminal one to two, one to three and two to three. On a single-phase starter only one reading is taken.

RPM READING INSTRUMENTS

Rpm instruments in testing and balancing are used to measure the actual revolutions per minute of fans, pumps and motors in order to determine if they meet design or not and to evaluate the fan performance. It is the second fan performance characteristic needed to evaluate or change the fan performance.

The most common types of tachometers used in balancing are:

Direct contact **manually timed** rpm counters
Direct contact **automatically timed** rpm counters
Direct contact **instant reading** digital counters
Direct contact **instant reading spring loaded** dial tachometers
Photo tachometers
Combination direct contact – photo tachometer with digital readout

Direct Contact Manual Rpm Counters

Direct contact manually timed revolution counters are small and inexpensive rpm reading instruments. They can be carried in your pocket and thus be easily available at any time.

Timed readings must be taken. The rubber tip of the revolution counter is applied to the end of the rotating shaft. It can turn in either direction, clockwise or counter-clockwise.

Direct Contact Automatic Timed Counters

Automatically timed direct contact counters are faster and easier to use but also more expensive. They incorporate a combination revolution counter and precision stop watch in one unit, and are self-timing and self-zeroing.

Direct Contact Digital Counter

For the fastest and easiest readings with direct contact rpm counters, use the digital model. No timing is required with its instantaneous readout and no starting and ending counts have to be written down and subtracted.

Direct Contact Spring-Loaded Tachometer

Direct contact spring-loaded instantaneous reading counters are like the speedometer of a car, but have a degree of inaccuracy of plus or minus 25 rpm and are best used for spot checking.

Photo Tachometers

Photo tachometers read rpm directly off a reflective tape on the pulley or shaft without contact. They are excellent where space conditions restrict reaching the end of the shaft with a direct contact counter and are a safer means of taking rpm readings that direct contact types. The light beam of the photo tachometer can be directed toward the shaft or pulley from any available vantage point.

To use a photo tachometer, stick a light colored reflective tape on the shaft or pulley. Aim the light beam at marker at a slight angle not more than 30 inches away.

Press the button to actuate the meter. The photo electric cell will respond to the slightest reflective light change on the rotating object and transmit it to the built-in pulse triggered computer which counts the light change rate and computes the rpm. Read rpm directly on scale.

Combination Direct Contact And Photo Tachometer with Memory

Probably the most practical and versatile rpm reading instrument available for air balancing is the combination direct contact and non-contact reading photo tachometer with a microprocessor.

It not only allows you to take either a direct contact reading, where accessible, and a non-contact reading where the fan shaft is non-accessible, but the memory recalls the last reading, minimum/maximum and average readings.

STATIC PRESSURE READING INSTRUMENTS

After the electrical and rpm readings are completed, two sets of pressure type readings are required, the fan suction and discharge pressures and the pressure drops across filters and coils.

This information, along with the electrical and rpm data, provides the third and fourth fan performance characteristics required.

Static pressure reading instruments are used for:

- Fan suction and discharge pressures
- Pressure drops across major suction side components
- Pressures in ducts and plenums
- Trouble-shooting pressure drops across bad duct fittings, reheat coils, dampers, etc.
- Reading outside atmospheric pressure

Pressure readings are taken with the following instruments:

- Liquid Manometers, Inclined, Vertical, U-Tube
- Mechanical Pressure Gauges
- Pitot Tubes
- Velometers
- Electro-Manometers

The manometers and gauges are used in conjunction with sensors such as pitot tubes, metal and rubber tubes and various static pressure sensors.

Liquid manometers, velometers and electro-manometers are primarily used for air flow readings and secondarily for static pressure readings. Mechanical pressure gauges are used more extensively for static pressure readings rather than the above instruments and will be covered in this section. The duct air flow readings types will be covered in the next section.

MECHANICAL PRESSURE GAUGES

Liquid draft gauges are excellent instruments for velocity pressure readings in ducts, but are not always very practical for the many static pressure readings that are required in balancing.

For reading static pressures the mechanical magnehelic gauge is easy to set up and far handier than the liquid gauges. They are used with various sensors, pitot tubes, strength tubes, static sensors, rubber tubes, and so on.

The magnehelic is a dry type manometer and has internal bellows for sensing pressure which are connected to the scale needle.

The magnehelic gauge has many uses in the balancing and maintenance of air distribution systems. It can be used for reading:

- Pressure drops across filters, coils, dampers, duct runs in balancing and trouble shooting
- Fan suction and discharge pressures
- Pressures at the inlets of constant and variable air volume terminal boxes
- Flow or pressure measuring stations.

(See Figures 9-5 and 9-7.)

Magnehelics come in a variety of ranges of inches of water column such as one quarter inch, half inch, 1, 2, 3, 5, 10 inches, etc. There are also combination pressure and air velocity reading models available where the pressure is read on the top of the scale and the fpm directly below.

1. Select the appropriate scale gauge for the reading. Use the smallest scale gauge possible for maximum accuracy.

2. Zero for a vertical or horizontal position reading.

3. Connect hoses to either one or both ports on the upper left side and then to a straight metal tube, pitot tube or static probe. For a single positive static reading, such as the discharge side of a fan, connect to the upper "high pressure" port. For a negative reading, such as on the suction side of the fan or in the exhaust duct, connect to the lower "low pressure" port.

ELECTRO-MANOMETERS

Alnor introduced two new electronic micro-manometers to the testing and balancing world in the past two years.

Compu-Flow Electro-Manometer

The first model came out late in 1985 and is the Compu-Flow Electro-Manometer available with an attached printer. This instrument is used in conjunction with a pitot tube for taking traverses velocity readings in ducts to determine air flow. Two rubber hoses are attached to the terminals on the bottom of the electro-manometer and to the pitot tube in a normal manner.

The multiple traverse readings are recorded, added up and averaged automatically. With the printer, both the FPM and velocity pressure in inches of water gauge are printed as well as the average velocity. Either metric and English units can be read.

Positive and negative pressures can be read as well as differential pressures.

There is no need to go through the process of venting, leveling, zeroing and finding a flat steady surface as is required with a liquid manometer. Also, the Compu-Flow can be moved around and be held in any position without resetting it.

Eco Electro-Manometer

Just recently Alnor made their second electro-manometer available, a smaller version of the Compu-Flow model above the ECO model 530. It reads positive and negative pressures from –1 to 10 inches of water and air velocities between 360 and 9990 fpm. It does not record readings, average velocities or print them out, but it's about half the price of the full version Compu-Flow and does the job.

Mu-Tube Manometers

U-Tube manometers are inexpensive liquid instruments for measuring positive, negative or differential pressures with a high degree of accuracy across air distribution system components, natural gas pressures and hydronic balancing.

Since water is used in U-tubes it gives you the full-scale measurement in inches of water column.

INSTRUMENTS USED FOR AIR FLOW READINGS IN DUCTS

To read the total air volume being delivered by the fan, a pitot tube traverse is made in the discharge duct with a manometer and pitot tube. For low velocity readings a quarter-inch liquid manometer or electro-manometer is used along with the pitot tube. For medium and high velocity readings in ducts, a vertical manometer or a higher scale inclined is required, if a liquid gauge is used.

Inclined Liquid Manometer

The 1/4-inch inclined gauge is used for taking on accurate air velocity traverse with the pitot tube in low pressure ducts in a range between 40 and 2000 fpm. The air velocity is then multiplied times the duct cross sectional area to determine the air volume.

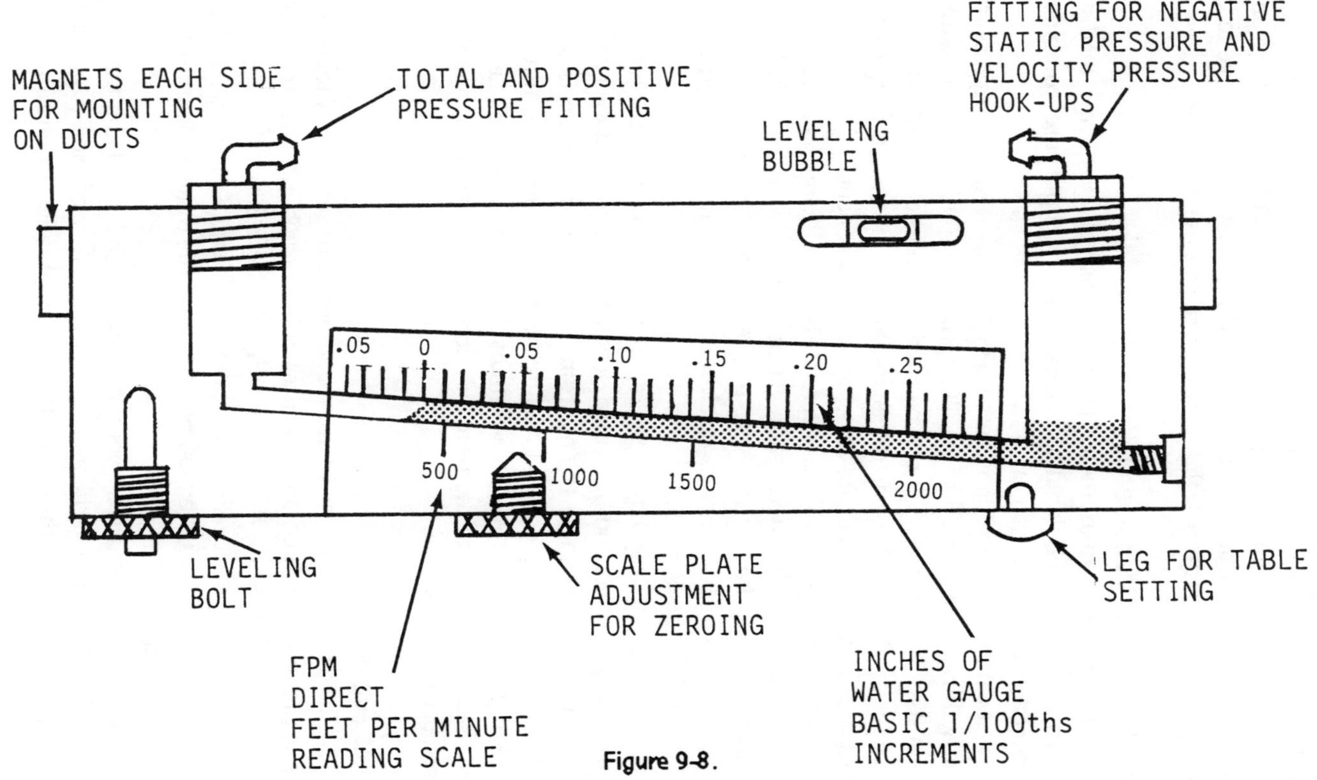
MAGNETS EACH SIDE FOR MOUNTING ON DUCTS
TOTAL AND POSITIVE PRESSURE FITTING
LEVELING BUBBLE
FITTING FOR NEGATIVE STATIC PRESSURE AND VELOCITY PRESSURE HOOK-UPS
.05
0
.05
.10
.15
.20
.25
500
1000
1500
2000
LEVELING BOLT
SCALE PLATE ADJUSTMENT FOR ZEROING
LEG FOR TABLE SETTING
FPM DIRECT FEET PER MINUTE READING SCALE
INCHES OF WATER GAUGE BASIC 1/100ths INCREMENTS

Figure 9-8.

The Dwyer 115-AV has two scales. The top one, in black, is in inches of water and the bottom one, in red, is in feet per minute.

To set up a liquid gauge, they must be vented, leveled and zeroed and cannot be moved once set up; otherwise, they have to be reset. They also require a flat, steady surface to set them on.

Vertical Liquid Manometers

The vertical manometer is an extensively used air reading instru-ment by professional balancers. It is a dual purpose, combined inclined and vertical manometer.

It can be used for high pressure systems as well as low pressure, and will read air velocities from 400 fpm to 10,000 fpm and pressures from 0 to 10 inches w.g. It can be used in conjunction with the pitot tube for air flow readings in ducts, for fan suction and discharge static pressures and pressure drops across system components.

The upper part of the Dwyer 400-10 is an inclined gauge with a black scale over the fluid reading from 0 to 1 inch and a red scale under-neath reading from 500 fpm to 4000 fpm. The vertical column reads from 1 to 10 inches in black to the right and to the left in red from 4000 to 12,600 fpm.

Like the inclined liquid gauge, the vertical model must rest on a flat, steady surface, and be vented, leveled and zeroed at every use.

Connect the rubber tubing from the sensor to both ports on the vertical gauge for velocity pressure reading. Connect to left-hand side of gauge for positive, above-atmospheric pressure static pressure readings, or connect to the right side for negative under-atmospheric pressure readings.

Pitot Tubes

The pitot tube is used with manometers to sense both total and static pressures in ducts and plenums. It is inserted into the air stream parallel to, and facing the air flow. See Figure 9-6.

Pitot tubes are double wall tubes constructed of stainless steel. The inner tube senses the total pressure and the outer tube senses the static pressure.

The total pressure enters the center of the tip of the main leg facing the air stream and comes out of the center of the bottom. The static pressure enters through a ring of small holes around the main leg, travels through the outer tube and out of the short leg near the bottom.

Static and total pressures can be read separately if required with a single hook-up to either the static or total pressure legs.

Velocity pressure readings are read by hooking up simultaneously to both the static and total pressure legs and to both sides of the manometer at the same time. The static pressure leg, which is normally the lower pressure, is hooked up to the right side of a liquid manometer and the total pressure which is usually the greater pressure is hooked up to the left side of the manometer. The two forces internally subtract themselves in the gauges resulting in a velocity pressure reading.

Pitot Tube Traverses

To determine an accurate average air velocity in a duct, a number of readings must be taken over the cross sectional area of air flow and averaged out.

The basic approach in determining the number of points to read is to divide the cross sectional area of the duct into equal size areas and take readings in the center of each. This may result in anywhere from 3 to 12 points being read horizontally across the duct and anywhere from 2 to 12 rows vertically depending on the size of the duct. Spacing between points of readings may be between 4 inches and 12 inches, which again is relative to the duct size.

Generally the first and last points in a row or column are anywhere from 1 to 6 inches in from the walls of the duct. The procedure for determining the number of points in a row or column is to divide the width or depth of the duct by the recommended spacing and split the remainder at the walls of the ducts.

Figure 9-6 shows a 15-point traverse of a 40" x 18" duct. Dividing the recommended 8-inch spacing into the 40-inch width gives four 8-inch spaces, with 8 inches remaining to be divided in half for the spaces on the sides, which makes the spaces 4 inches each. The depth of the duct is treated in a similar manner.

After the traverse is laid out, determine whether to take the readings through the bottom of side of the duct, and drill 3/8-inch-diameter holes in the proper locations. Mark the depth of insertion for each point to be read on the pitot tube with take or a marking pen.

Set up your manometer, connect the hoses, insert the tube through the first hole and read each point. If your gauge has fpm on the scale, read it directly and record the fpm. If your gauge only reads velocity pressure in inches of water gauge, read and record them on your form.

Every point in the traverse must be recorded even if the velocity is zero!

After completing and recording the readings for each point add up the fpm and divide by the number of points for the average velocity. If velocity pressures are recorded in inches of water they must be converted to fpm first before adding them together; otherwise, the average fpm will be erroneous. Do not add the water gauge readings together directly, **convert first.**

READING AIR FLOW AT OUTLETS

After all the equipment is checked out, the amps read, the fan rpm taken, the discharge duct traversed for total air flow, the fan static pressure and component pressure drops measured, you are ready to balance the outlets and individual duct runs.

Another set of instruments are brought into play here for this purpose which are categorized as outlet air flow reading instruments and used for:

- Exhaust Grilles and Registers
- Supply Diffusers
- Exhaust Hoods
- Duct Openings
- Air Flow Through Filters, Coils, Louvers

Various Types of Outlet Air Flow Reading Instruments Available

- Air Volume Hood with Modified Velometer Meter
- Junior Size Air Volume Hood with Jr. Velometer
- Air Volume Hood with Micro-Based Multi-Use Meter
- Swinging Vane Velometer

- Mechanical Rotating Vane Anemometer
- Micro-Based Digitized Rotating Vane Anemometer
- Pocket Type Rotating Vane Anemometer
- Thermal Anemometers

Air Volume Hoods

An air volume hood completely engulfs a diffuser or grille, forces all the air to flow down through it and measures it directly in CFM with one reading in a few seconds. Air volume hoods save a great deal of time and hassle involved with ak factors, multiple readings, averaging and converting fpm to CFM which is the case with velometers. They can be used effectively to determine air flows at supply diffusers and exhaust grilles by practically anyone regardless of their balancing experience.

It consists of a nylon cloth top, and a metal bottom housing with a differential air pressure measuring manifold in it which is hooked up to either a velometer or electro-manometer mounted on the outside. A CFM range selector next to the metal compensates for variable backup pressures.

There are various size hoods available depending on the size outlets being read, 2 ft x 2 ft, 1 x 4, 2 x 4, 1 x 5 and 3 x 3.

Air Volume Hood with Modified Velometer Meter

Alnor manufactures two hoods today. One is a full-size air volume hood, for reading air flows at outlets. It is a fast, easy-to-use, light weight, quick to set up instrument and comes with a carrying case.

It has a blue fabric hood, metal base, sensor, manifold, range selector and low flow screen accessory. There is a full range of five size hoods available.

A modified velometer is mounted on the metal base and hooked up to the manifold for direct supply and exhaust CFM readings.

New Junior Air Volume Hood

Alnor has also just released a low-priced Junior-Type air volume hood for more limited spot checking, trouble shooting, ease of portability, etc. for those who are not that heavily involved in testing and balancing and who do not have to cover the full range of all size outlets.

It comes with two hoods, 24" x 24" and 16" x 16" sizes, which are shorter in length than the above regular hoods. It consists of a light-weight, metal bottom, sensing manifold, easy-to-install low flow screen, CFM range selector and the smaller Velometer Jr. instead of the full-size velometer.

Air Volume Hood With Micro-Based Multi-Use Meter

Shortridge Instruments has a new air volume hood with a multi-use electronic micromanometer which not only reads air flow, but three other elements, air velocity, air pressure and temperature.

It reads a complete range of all flows, velocities, pressures and temperatures from laminar flows in clean rooms to high pressure systems.

The meter can be removed from the flow hood and used separately for other readings.

The Velometer

The velometer is an instantaneous direct reading instrument which reads air velocities at diffusers and grills in fpm and pressures in ducts in inches of water column. It required a four-point velocity reading, mathematical averaging, and looking up ak factors in order to calculate the CFM flow.

Its main use over the past 10 or 15 years has been to read supply air velocities at ceiling diffusers. It can also do several other things such as reading exhaust velocities at grilles, hoods and duct openings, air currents in open spaces, velocities at filters, coils and louvers and pitot tube traverses and static pressures.

It was originally developed by a group of testing and balancing pioneers in the Chicago Sheet Metal Contractors Association in conjunction with Alnor.

Two Types of Alnor Velometers

There are two types of Alnor velometers, the old swinging vane series 3002 models which are no longer manufactured, and the new 6000P series, a greatly improved version with a taut band added to the vanes so that it can be held in either a horizontal or vertical position.

The most commonly used models in HVAC balancing are the heating and air conditioning set 6000BP and the all-purpose set 6000AP. The kits are composed of the velometer itself, a diffuser probe which can be used for either supply or exhaust fpm readings, range selectors, a low flow probe, hoses, a pitot tube and static probes.

The HVAC 6000BP model reads from 0-2500 fpm with the 6070P probe and reads from 0-300 fpm with the low flow probe.

To set up the velometer attach two rubber hoses to ports on the back of the instrument and to the terminals on the bottom of the range selector.

Insert the diffuser probe into the range selector. Set the range selector to the correct velocity range, look in the manufacturer's ak factor book for the number of readings to be taken, the location of the points of readings, the proper position of the probe–**and follow precisely!**

Generally readings on round and square ceiling diffusers are taken at four points. The probe is set on top of the lip of the outer most diffuser cone and held perpendicular. The four readings are added together and divided by four for the average velocity.

Mechanical Rotating Vane

Anemometers are used by HVAC balancers to read air velocities in feet per minute at grilles, louvers, exhaust hood openings, duct openings, at louvers and at filters and coils in sheet metal housings.

The standard anemometer is four inches in diameter, has a rotating vane wheel which operates on the windmill principle, a large round feet-per-minute dial, two smaller dials for reading hundreds and thousands, a jeweled gear box, an on-off lever at the top and a zeroing level on the side.

Timed readings must be taken, usually for half a minute or a full minute. The revolutions of the rotating vanes correspond to the feet-per-minute air speed traveling through the instrument. Accuracy is generally about 2 or 3 percent. Rotating vane anemometers need to be correctly calibrated and may require correction factors for non-linear readings.

Micro-Based Digital Rotating Vane Anemometer

The mechanical rotating vane anemometer, which requires timed readings and involves difficulty in reading the dial scales, has evolved into a remote micro-based digital read-out station which gives highly accurate, easy-to-read digital readings in a few seconds.

It is an excellent answer for readings of grilles where there is inadequate space for a flow hood or where the grilles are very large. It can work well with exhaust hood openings, reading air flows through large banks of filters or coils, and laminar flow readings in clean rooms or surgical areas.

It is easily portable, has a memory for average speed, maximum and minimum speeds.

Several firms offer digital rotating vane anemometers such as: Pacer Industries in Chippewa Falls, Wisconsin/ Testoterm, Inc. in Nashville, Tennessee.

Pocket Type Digital Rotating Vane Anemometer

Another excellent enhancement to the mechanical vane anemometers is the pocket type which has the digital meter built right in with the rotating vane. It is low priced and very easy to carry around.

Thermal and Hot Wire Anemometers

Anemometers with electrical sensing elements are excellent for quick, accurate and highly sensitive measurements of low air speeds such as:

- Air currents in rooms, drafts and terminal velocities
- Flows in exhaust hoods
- Air flows in clean rooms and surgical spaces
- Velocities at points of origin in industrial exhaust work
- Determining velocity distribution around HVAC equipment

They can be used for spot checking both diffuser velocities and pitot traverses in ducts. Some models can also measure temperature and air pressure as well as air velocity.

There are two basic types of electrical anemometers, thermal anemometers and hot wire anemometers. They are similar in that with both of them air passes over an electrically charged sensing element in a probe. The probe must be exposed directly to the air stream and be protected so the element doesn't touch anything.

Thermal anemometers work on the principle of the cooling capacity of air. When a heated probe is placed in an air stream it will be cooled and the magnitude of the cooling is related to a direct velocity reading.

The hot wire type anemometer has a sensing element which is heated electrically and kept at a constant temperature. When the element is exposed to the air stream either the temperature difference between the element and air stream is measured, or the electrical current, which is needed to maintain the constant temperature, is measured and calibrated to the velocity of air flow.

Kurz Portable Air Velocity Meter

The Kurz Portable Air Velocity Meter Model 440 is a rugged thermal anemometer that performs many testing operations.

The Model 440 has three velocity ranges of 0 to 300 fpm, 0 to 1250 fpm and 9 to 2500 fpm, and static pressure range of 0 to 3 inches of water.

It has an accuracy of plus/minus 3 fpm in the lower velocity ranges and plus/minus 2 percent accuracy for each full-scale range.

DUCTWORK LEAK TESTING

Unsealed low pressure ductwork with cleat connections can leak anywhere from 5 to 15 percent without necessarily adversely affecting the performance of the system. Medium or high pressure ductwork, on the other hand, might only leak initially 1, 2, or 3 percent, depending on how well it was sealed, but lose its ability to function with leakage over 1 percent.

Medium and high pressure ductwork must be properly sealed at connections, along lengthwise seams, and be able to hold duct pressures, or the terminal units won't operate. Duct connections are responsible for the major part of duct leakage and must be sealed in the field.

Leaks also cause other problems besides pressure losses such as air noises, hissing, ballooning cement at seams, blown insulation and CFM losses. Consequently the ductwork must be leak tested in order to locate and seal faults.

There are numerous existing medium and high pressure systems where there is, unknowingly, excessive leakage, causing terminal units not to operate properly, or not deliver any air at all. There are systems that are not under control and never were.

Maximum Allowable Leakage

The maximum total allowable CFM leakage as set by ASHRAE and SMACNA is between 1/2 and 1 percent of the system total. One percent is generally sufficient. This means that a 30,000 CFM system, as a total, may leak 1 percent times 30,000 which equals 300 CFM.

Testing Pressures

The system or segment to be leak tested is pumped up, that is, filled with air, and pressurized to either 2 inches over or 25 percent over the system design static pressure. Hence a 6-inch system is tested at 8 inches.

Equipment Required

The equipment required for leak testing, as shown in Figure 9-9, is as follows:

a. High-speed direct drive vacuum type fan, 3600 rpm, 75-100 CFM. Cadillac is one manufacturer.

b. Calibrated 3 or 4 inches diameter orifice testing tube with straighteners and orifice plate. Pressure drop across orifice plate corresponds to CFM being lost.

c. Two "U" types with tubing for hookups.

d. Flexible duct tubing to connected orifice tube to duct run to be tested.

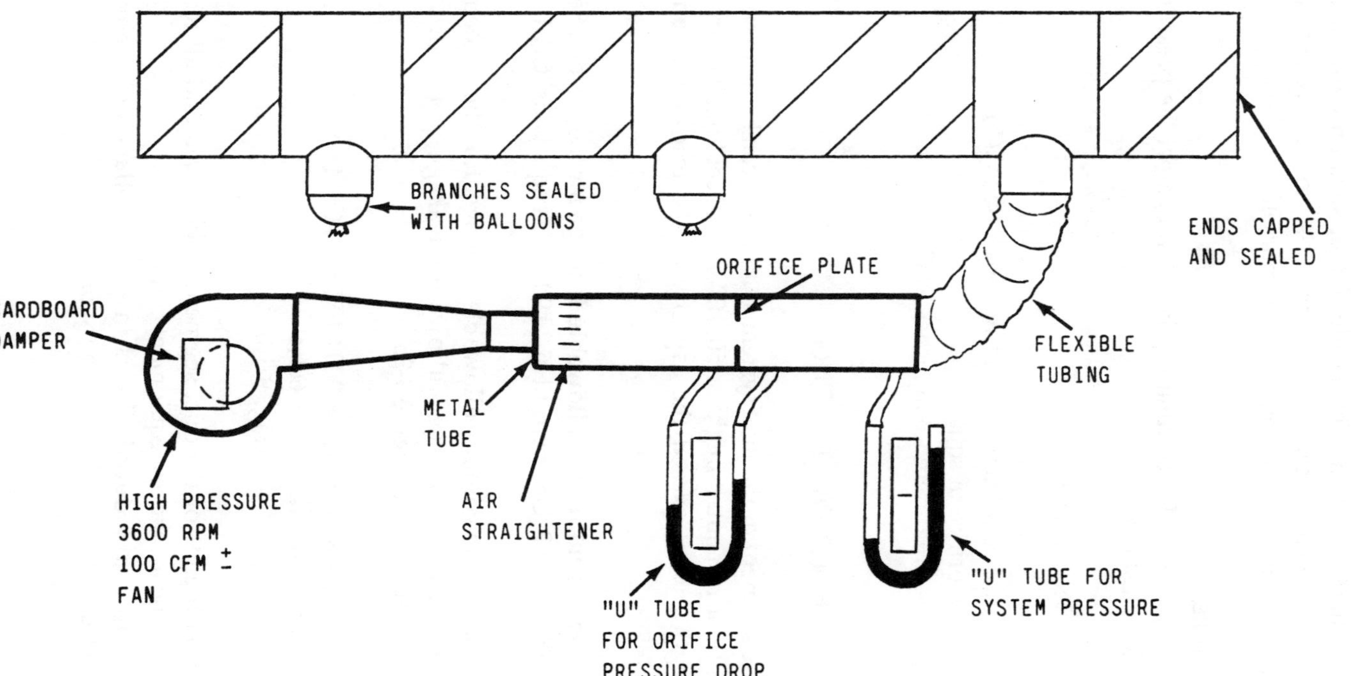

Figure 9-9.
Leak testing set up for checking leakage in medium or high pressure duct.

United Sheet Metal in Westerville, Ohio, sells an entire leak testing rig with a certified calibration chart. It is more convenient to purchase from them than to assemble a rig yourself.

Orifice Formula

The formula used for calculating the CFM against the pressure drop in the leak test curve is:

CFM = 4005 x C x A x P

where

C = orifice coefficient (generally about .60 to .65)
A = area orifice in square feet
P = pressure difference in inches W.G.

PROCEDURE FOR LEAK TESTING

1. **Test one segment of the duct system at a time.** It can be main run, riser or floor run out. It might be 50, 100 or 150 feet long, however the length cannot exceed the capacity of the test blower and orifice tube.

2. **Determine the minimum allowable leakage** for the system. See example Figure 9-2. One percent of 20,000 CFM equals 200 CFM.

3. Determine the maximum allowable leakage for the system. Work on basis of average allowable leakage per foot of ductwork. See example Figure 9-2.

$$\frac{\text{Maximum System Leakage}}{\text{Feet of ductwork}} = \frac{200\text{ CFM}}{600\text{ ft}} = .33\text{ CFM/ft}$$

(Fig. 9-10)

4. Seal off both ends of the duct segment to be tested and all tees except the one for the test equipment connection.
   Blowing up heavy duty balloons in round collars and sealing with duct tape seals well and is very fast.

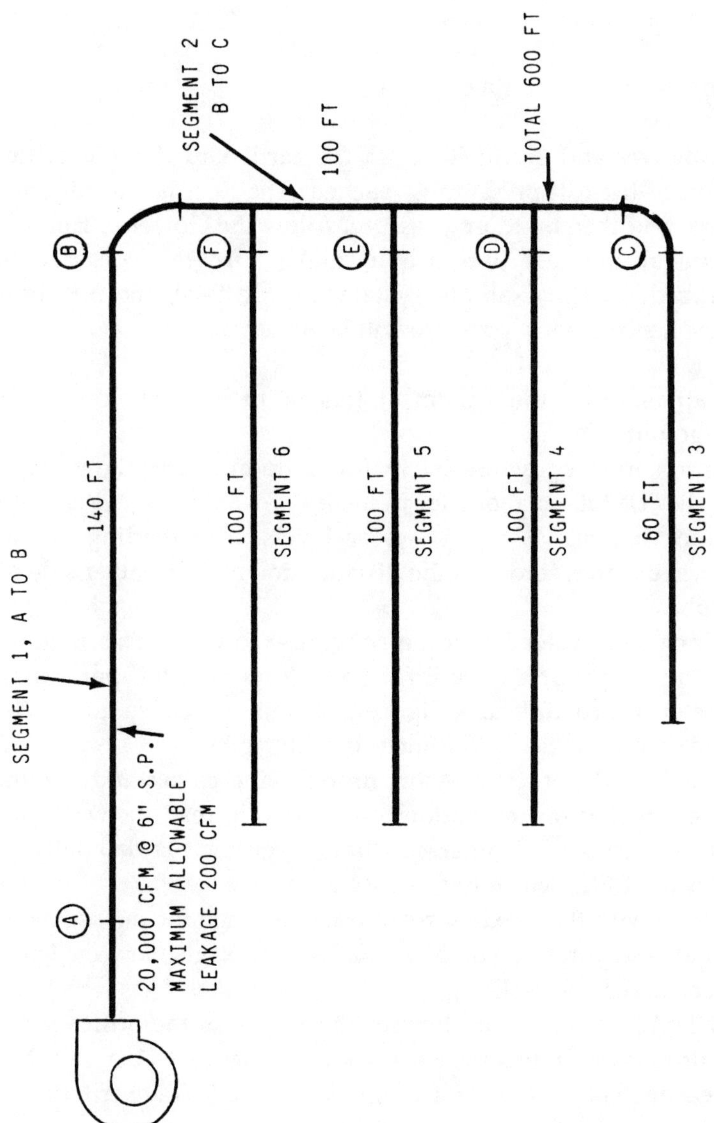
A
B
C
D
E
F
SEGMENT 1, A TO B
140 FT
20,000 CFM @ 6" S.P.
MAXIMUM ALLOWABLE
LEAKAGE 200 CFM
SEGMENT 2
B TO C
100 FT
TOTAL 600 FT
100 FT
SEGMENT 6
100 FT
SEGMENT 5
100 FT
SEGMENT 4
60 FT
SEGMENT 3

Figure 9-10.

Other methods of sealing openings are with food containers, plastic bags and metal caps.

5. Set up and connect testing rig.

6. Start blower and gradually open the cardboard damper at the inlet until the test pressure is reached. The duct has to fill with air first and then build up pressure. Allow the blower to run one minute at the test pressure to make sure the pressure is stabilized. In the example system, Figure 9-00, the S.P. is 6 inches, hence the test pressure will be 8 inches.

7. The air entering the system at this point is equal to the air leakage out.
   The pressure drop across the orifice plate measures the leakage with the "U" tube manometer in inches of water gauge. Read the distance between the two water levels. The reading is the distance up from zero plus the distance down, in the other side of the tube.
   The inches of water has to be converted to CFM. This is done with an appropriate calibration curve for the particular diameter orifice and tube. Figure 9-00 shows a curve for a 7/8-inch-diameter orifice and 3-inch-diameter tube.
   Let's assume the orifice pressure drop is 6 inches to start off with. Locate 6 inches on the bottom of the graph, read up vertically until the curve line is intersected and then read horizontally to left for the CFM, which in this case is 25 CFM.
   Compare with the maximum allowable leakage for the segment, 20 CFM, and you find you have excessive leakage. Now the leaks must be found and sealed.
   Now let us assume that after sealing the leaks the orifice pressure drop is down to 4 inches. Looking at Figure 9-11 shows that the leakage is now only 20 CFM, meeting the leakage criteria.

8. Walk the duct run listening for audible leaks and visually looking for holes, cracks, ballooning cement or bulging tape. Soap bubbles may be required if there are large numbers of small leaks. Mark the leaks.

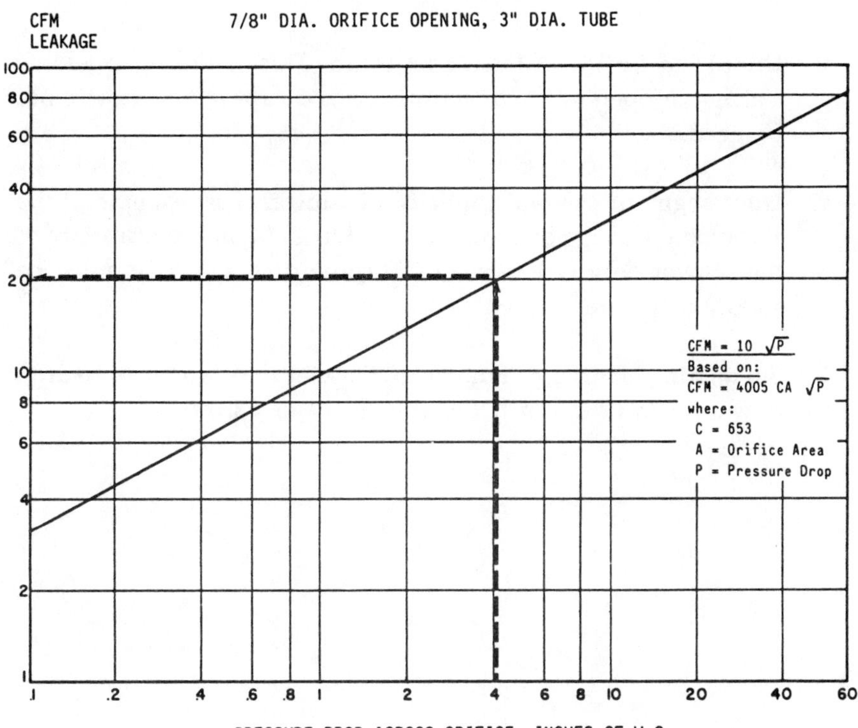

Figure 9-11

9. Turn off the test fan and seal the leaks. After the sealant has set, run the fan up to test pressure again and recheck leakage. If still excessive, check duct run again and reseal. If leakage is 1 percent or under, recheck duct run anyway to make sure there are no audible leaks remaining.

Trouble-Shooting

A. **Constant build-up of pressure in duct cannot be lowered to test pressure.** This occurs if the blower is pumping more air in than the duct can leak out and means that the duct is sealed very well.

B. **Unable to build up to test pressure.** This indicates excessive leakage or openings in ducts such as end caps left off or tees not plugged.

C. **Duct segment beyond capacity of blower.** For example, if the acceptable leakage on a duct run is 120 CFM and the capacity of the blower only is 100 or 75 CFM, the duct segments must be subdivided more.

D. **No reading on orifice manometer.** Stays at zero. No pressure drop means no air flow which means no leakage. **Hurray!**

Chapter 10
TROUBLE SHOOTING AIR DISTRIBUTION SYSTEMS

Problems with air distribution systems in commercial, industrial and institutional buildings stem from the multitude of components in systems, from huge fans to tiny transistors. Problems stem from designs, installation, equipment, controls, balancing, changes, maintenance, operations, etc.

Problems also stem from the uncontrollable variability of heating and cooling load conditions of the outside environment, and from the variable internal usages.

Problems exist because of the wear and tear on HVAC system components and the inevitability of human error.

> *A fan may be pumping out 30 or 40 percent less air because of dirty components or bastard ductwork fittings. There may be double the air needed near the fan and only half at the end of the duct run due to system imbalance. The perimeter of the building and the area near the floor may be too cold, while the interior of the building and the area near the ceiling level may be too hot.*

TYPICAL PROBLEMS AND COMPLAINTS

The most frequent problems and complaints with air distribution systems are as follows:

- It's **too hot**
- It's **too cold**

- It's **drafty**
- There's **too much air**
- There's **not enough air**
- It's **too noisy**
- **Energy costs are too high**
- **Operation problems and excessive maintenance**
- **Excessive air distribution system resistance**
- **System out of balance**
- **Fan problems** with wheels, motors, belts, bearings freeze

GENERAL CAUSES OF AIR DISTRIBUTION PROBLEMS

HVAC problems fall in four general areas:

Cooling System
Heating System
Air System
Control System

Problems in any of the above four areas can be induced by the following conditions and situations:

1. Incorrect flow rates and delivery temperatures of air or fluids.
2. Design errors in the heating/cooling load, flow rates, ductwork or piping sizing and resistance calculations.
3. Occupant changes in the usages of spaces, internal heat generating components, of HVAC systems and architectural remodeling.
4. HVAC equipment is not set, adjusted, hooked up or functioning properly. A fan running at the wrong rpm of backwards. The wheel and cone misaligned, the wrong type of wheel or the wheel installed backwards.
5. Broken, worn or malfunctioning HVAC equipment.
6. Dirty or clogged HVAC components, filters, coils, turning vanes, dampers, etc.
7. Imbalanced system.
8. Outside air and mixed air may not be under proper control.

9. Poor or inadequate temperature control zoning. Zoning not compensating for shifting sun loads, internal loads, occupancy etc. Poor mixture of spaces on a zone.
10. Drafts, poor air velocity and temperature gradients, stratification in spaces.
11. Controls for dampers, valves, thermostats, pressure and flow sensors, not set correctly, located effectively or malfunctioning.
12. Poor operation and maintenance.
13. Type and capacity of HVAC equipment chosen in the design process incorrect.

EXTRA ENERGY COSTS DUE TO AIR DISTRIBUTION PROBLEMS

1. Excessive resistance in air distribution system caused by dirty or clogged components, air imbalance, high resistance filters, dampers closed, poor ductwork design or incorrect installation generate geometrically higher energy costs.
   An extra 1/2" static pressure or resistance causes the fan to draw one extra brake horsepower.
   Hence, the additional operating costs for each additional BHP being drawn from motor at 8¢ per kWh are:
   Extra $700 per year, full time
   Extra $350 per year, half time

2. Too much air from fan. A fan pumping out excessive air flow, for example 20 percent too much, is expending 50 percent more horsepower than it needs to.
   This means if a fan is delivering 12,000 CFM instead of 10,000 CFM, it may be drawing 7-1/2 BHP instead of the rated 5 BHP.
   Hence at 8¢ per kWh an extra 2-1/2 BHP costs:
   Extra $1,750 per year, full time
   Extra $875 per year, half time

CHECKING PROCEDURE

If the space or supply outlet temperature are off drastically, overall in a system, check the operation of the cooling, heating and control systems first. If they appear to be operating properly, proceed with checking the air distribution systems.

An effective procedure for trouble shooting air distribution system problems is as follows:

1. Check the thermostat for setting, operation, location and calibration first.
2. Supply outlet air flows and discharge temperatures.
3. Return air and exhaust grille air flows and intake temperatures.
4. Air flows in branch and main duct runs.
5. VAV or CAV terminal boxes for enough pressure at inlet to operate, functioning correctly with stat, controls correct on the box.
6. Inspect ductwork in ceiling spaces for missing end caps, disconnected ductwork, wrong duct connections, excessive leakage at duct connections.
7. Check fan total air flow, rpm, suction and discharge pressures and motor amp draw.
8. Check filters and air handling unit coils if dirty or clogged. Check for velocity or temperature stratification of air through coil.
9. Check if outside, return and exhaust air control dampers are operating properly.
10. Budget check approximate actual heating/cooling load and CFM required for the whole system. Also check approximate heating/cooling load and CFM required for individual problem spaces. Compare with design requirements and actual readings.
11. Make sure the return air and exhaust fans serving the spaces are turned on, are operating properly and at correct capacity.
12. Check for negative and positive pressures in spaces. Check for diversity between supply, RA, exhaust fan volume and makeup air requirements, etc.
13. Check if the system is balanced.
14. Are there noise problems at outlets or at fan? Check for vibrating ducts, pulsations, throbbing, etc.

Checking Procedure

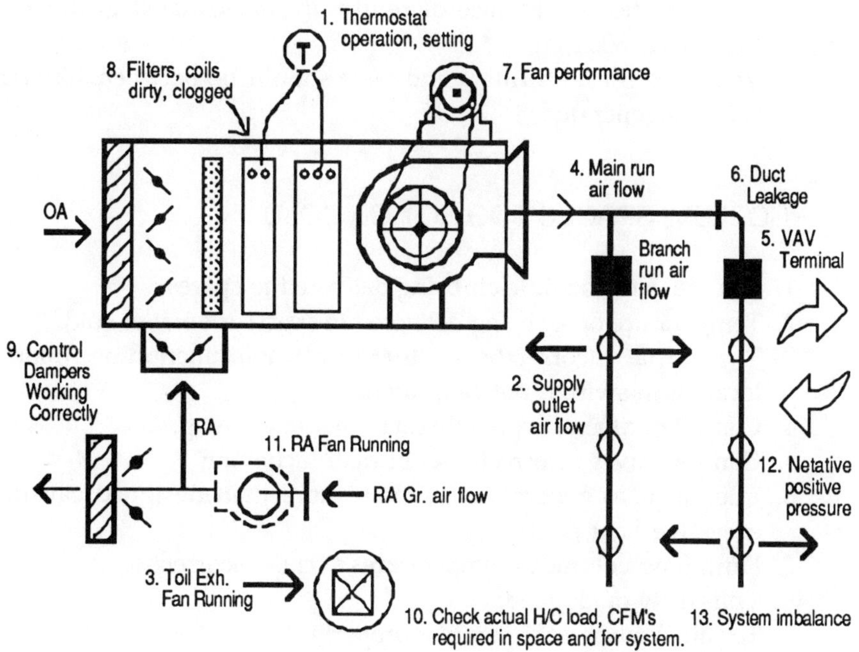

Figure 10-1.

TYPICAL AIR SPEEDS AND TEMPERATURES

Having a general idea of what air speeds and temperatures generally should be, greatly facilitates the trouble-shooting process. The following are common numbers to gauge yourself against when taking these readings:

1. Cooling supply air typically is designed at 55°F.
2. Heating supply air may range from 90°F to 130°F.
3. Low velocity trunk runs 1200 to 1600 fpm; branch runs, 600 to 900 fpm.
4. Medium pressure systems, trunk runs 2000 to 3000 fpm.

5. Discharge velocities at the face of ceiling supply diffusers, 600 to 1800 fpm depending on the required throw, noise criteria, etc.
6. Air velocities at the face of return air and exhaust grilles run from 200 to 800 fpm.
7. Velocities through filters and coils should be between 400 and 600 fpm generally.

PROBLEM: SPACES TOO HOT OR COLD

1. Too much or too little air being delivered to spaces.
2. Temperature of air being delivered to spaces too hot or cold.
3. Thermostat incorrectly set, turned off, malfunctioning, poorly located, miswired, not calibrated.
4. Control dampers incorrectly set or malfunctioning.
5. Control valves incorrectly set or malfunctioning.
6. Internal heat generation in room more than designed causing excessive heat gain.
7. Fluid flow volume or temperatures at coils incorrect.
8. Coils dirty or clogged.
9. Terminal boxes not operating properly.
10. Errors in heating or cooling load calculations.
11. Poor zone temperature control built into system not accommodating shifting sun loads, occupancy and so on. Poor mixture of spaces on the same zone.
12. System imbalanced. Too much or too little air in branch ducts and outlets near fan, too little at end of duct runs.
13. System misdesigned. Duct sizing incorrect, pressure calculations off.
14. Occupant changes. HVAC system, architectural, internal heat generation.

PROBLEM: TOO MUCH AIR

1. Less resistance in system than designed. Ductwork oversized, less ductwork than design, system components less resistance than design.

2. Not all the system components installed.
3. Access doors open on intake plenum.
4. Coil bypass dampers open along with face damper.
5. Fan oversized, rpm high, wrong type, system resistance less than design.
6. Over-design of heating/cooling load or air volume.
7. Windows and doors open in spaces.
8. Control dampers not set properly or malfunctioning.
9. Manual volume dampers open excessively.
10. Terminal units in medium or high pressure systems in open positions, wrong size, wrong controls, set wrong or malfunctioning.
11. Control of dampers, fans or terminal units not working or incorrectly set.
12. Excessive negative pressure in spaces. Too much direct exhaust air or return air being drawn out of spaces.
13. Grilles and ceiling diffusers not dampered down correctly, too large or wrong type.
14. Filters not installed at all or wrong type.
15. Coils wrong size or type.
16. Air density lower than design because it's warmer or at higher altitude.
17. Imbalance in air distribution system. Too much air in branches and outlets near fan and not enough in duct runs and outlets.
18. Poor temperature control zoning of system to accommodate changing heating/cooling loads during day or poor mixture of spaces on same zone.

PROBLEM: NOT ENOUGH AIR

1. Supply diffuser and grille dampers closed too much. Wrong size or type.
2. Too much resistance in air distribution system. Dirty components, under designed ductwork, excessive ductwork, bastard fitting, etc.
3. Space thermostat incorrectly set, located, calibrated, malfunctioning.

4. Excessive duct leakage. End caps missing, disconnected ducts, access doors open, excessive leakage at connections.
5. Dirty, blocked off, or clogged manual dampers, reheat coils, turning vanes, grilles, ceiling diffusers, fire dampers.
6. Fan undersized, rpm low, wrong type wheel, rotation backwards, wheel installed backwards, wheel and inlet cone misalignment or incorrect gap, cutoff plate missing. Manufacturer rating incorrect. Air starvation in suction side of fan.
7. Under-design of heating and cooling load or air volume.
8. Manual volume dampers closed too much.
9. Fire dampers accidentally shut.
10. Fire dampers set incorrectly or malfunctioning.
11. Filters dirty, clogged, blocked off, resistance rating too high, wrong type, wrong size.
12. Terminal units in medium or high pressure systems in closed position, wrong size or controls, set wrong, malfunctioning.
13. Excessive positive pressure in spaces. Not enough RA or exhaust air being drawn out.
14. Imbalance in air distribution system. Excessive dampering. Wrong procedures used in balancing. Poor design.
15. Poor temperature control zoning.
16. System incorrectly designed.

FAN RUNS, INSUFFICIENT AIR FLOW

1. Excessive resistance in system.
2. Wheel misalignment and improper wheel overlap and gap with the inlet cone can cause a sharp loss of air volume and cycling within the fan. (See Figure 10-2.)
3. Bad fan discharges can cause havoc with air quantities, statics, throbbing, pulsating, noises, duct wreckage. Poor and good discharges are illustrated in Figure 10-3.
4. Fan undersized
5. Fan rpm low
6. Low motor rpm
7. Wrong ratio of sheaves
8. Wheel rotation backwards

9. Wheel loose on shaft
10. Wrong type wheel
11. Wheel installed backwards

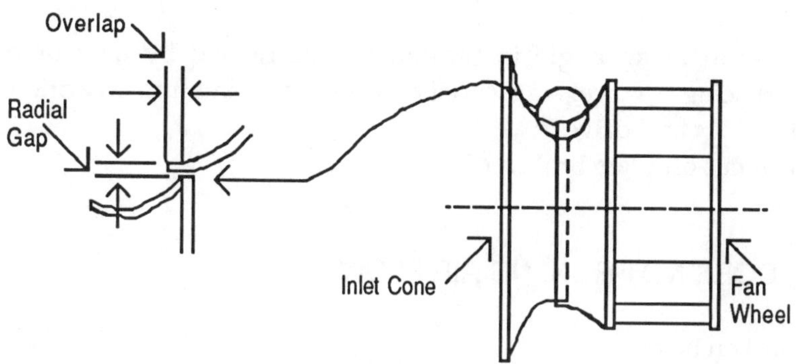

Figure 10-2

Improper gaps and overlaps of inlet come and wheel venturi can cause air cycling and drastic loss in fan volume.

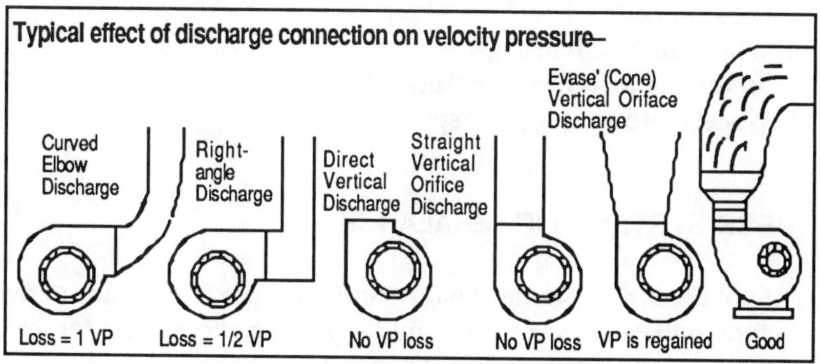

Figure 10-3.

FAN MAINTENANCE TROUBLE SHOOTING

1. Unbalanced wheels cause vibrations, noises and premature wearing of bearings and drives. Wheels have to be dynamically balanced and weights put on them for balance.

2. Misaligned or incorrectly tensioned belts on drives.
3. Wheel installed backward.
4. A warped shaft, which can be caused by heat or in removing a wheel, causes vibrations and premature wearing of bearings and drives.
5. A shaft at an angle in the fan can be caused by heat or by replacing a wheel, causes vibrations and premature wearing of bearings and drives.
6. Fan cut-off plate broken off.

FAN DOES NOT RUN, NO AIR FLOW

1. Broken belts
2. Blown fuses
3. Thermal overloads in starter kicked out
4. No power
5. Bearings froze
6. Motor defective
7. Loose pulleys
8. Wheel jammed housing
9. Motor overload heater kicked out
10. Wheel loose on shaft and spinning

CHECK SYSTEM FOR IMBALANCE

1. Spot check air volume at outlets farthest from fan, at ends of duct runs, outlets nearest fan and midway in the duct system for major discrepancies between actual and design CFM's.
2. spot check total air flows and static pressure in main duct runs.
3. Check total air flow at fan and suction and discharge pressures.

DRAFTS

1. The speed of air is too high when it hits the occupants, not in the 25 to 50 fpm comfort range.

2. The temperature of the air when it comes in contact with the occupants is too high or low in conjunction with the air speed. Should be in 68°F to 78°F range.
3. Too much air at diffusers or grilles.
4. Supply diffuser too small, deflection wrong or throw incorrect.
5. Doors or windows open.
6. Horizontal and vertical temperature stratification in spaces. Temperatures too high below ceiling or by outside walls and too low near floor or inside walls.
7. Horizontal and vertical air speed stratification and currents. Excessive air speeds near occupants, etc.

CHECK FOR POOR ZONING

1. Too many rooms on one zone.
2. Room that has thermostat for zone not representative of all spaces on zone.
3. Spaces in zone don't have homogeneous heating/cooling loads which correlate with time and conditions.
4. East, south, west, north solar load zones mixed on same system.
5. Perimeter and interior spaces mixed on same zone.

DEBUGGING VAV SYSTEMS

Even though VAV systems are the most promising and versatile HVAC systems around today, that doesn't mean that there aren't problems.

Design, equipment, controls, balancing, maintenance and operation problems exist and must be dealt with.

Let's look at ten of the most common bugs, at what is really happening, analyze the problem, determine the causes and come up with solutions.

1. **The problem: Glob of cold air** coming out of a ceiling diffuser sinks on occupants below. Air circulation poor in room.

This situation occurs on minimum flow when there is an insufficient flow and jet velocity to throw the air horizontally so the air hugs the ceiling along its course and reaches the extremities of the area.

Local velocities should be in the range of 25 to 45 fpm. Velocities under 25 fpm cause stagnation and over 45 or 50 fpm cause drafts. Other factors which affect comfort, such as relative humidity and surrounding radiant temperatures, may vary these velocity comfort ranges.

It is recommended that round diffusers should not exceed 300 to 350 CFM. Air distribution problems can occur at minimum flow rates also with round or square diffusers if they are sized at 400 CFM or more.

Volume reductions for round or square diffusers shouldn't exceed the 25 to 35 percent range.

Draft and air noise problems show up at maximum flow when diffusers are undersized. If you try to maintain throw and room circulation at reduced flow rates with large reductions of 5 to 60 percent on round or square diffusers, the system will probably generate unacceptable noise levels.

Outlets in general should be selected based on minimum CFM, outlet velocities, throw and noise levels.

2. **The problem: Maximum CFM not being held** by a terminal unit with high pressure at the inlet. A unit rated for 900 CFM is allowing 1100 CFM to flow through it. Or a terminal unit with low static pressure at the inlet won't operate at all.
   The box minimum inlet pressure may be 0.4" and only 0.2" is available. Possible causes of problems at the terminal units are:
   - Pressure control range of terminal too limited for system pressure variations.
   - Supply fan not controlled properly

- Return fan not controlled correctly
- Excessive system imbalance in the duct system at points between maximum and minimum flow
- Pressure-dependent unit selected instead of pressure-independent one
- Terminal may not be performing according to manufacturer's published performance data
- Terminal controls may be calibrating incorrectly, the maximum or minimum settings may be incorrect, they may be incapable of more accurate operation of they may be defective
- The actual diversity range may be greater or less than anticipated
- Ductwork poorly designed

There are a number of things that must be considered when selecting terminal units to resolve many of the above problems. Effective duct design involves having enough pressure in the duct at the intake of the terminal to not only overcome the pressure drop of the terminal itself, but also to overcome the low pressure discharge duct and end diffuser.
Larger medium and high pressure duct systems with velocities of over 2500 fpm and CFMs of over 20,000 should be designed with the static regain method. This is done so that roughly the same static pressure will exist at each duct and terminal branch and so the total pressure drop is the same to the end of each branch.
The equal friction method of duct design should be used for low pressure systems with velocities under 2500 fpm.

3. **The problem: Building is under excessive negative pressure.** This condition occurs both in the summer and winter, more severely in the winter.
The problem begins when the outside air quantity drops (along with a supply fan reduction) to the point that the outside air intake is less than the draw of the fixed exhaust fans. Let's say the outside air reduced to 2000 CFM in conjunction with the supply fan reduction, a deficit of CFM is created and hence a building negative pressure condition results.

The problem can be further aggravated in the winter by the following: Some building engineers blank off combustion air louvers in boiler rooms to prevent cold drafts and protect pipes from freezing. This lack of combustion air, combined with insufficient outside air, starves the ever-running fixed exhaust fans, creating yet a more disturbing negative pressure. The solution to this problem is to maintain a minimum amount of outside air with minimum outside air controls and adequate heating of combustion air.

One way to do this is to couple the outside air damper to the supply fan controller so that it tracks the operation of the fan vortex dampers. Or use a velocity controller, which measures the velocity of the incoming outside air and modulates the dampers to maintain minimum outside air.

4. **The problem: Oversize supply fan and motor.** Too much air and inefficient fan operation. A 40,000 CFM fan is selected when, due to diversity, a 30,000 CFM fan would have been a better choice. An oversize fan may not operate in its maximum efficiency range and may waste energy and increase operating costs.

   This problem is a result of:
   - Diversity not taken into proper consideration or ignored. Fan selected on total instantaneous load, not on maximum simultaneous load.
   - Over-calculation of system static pressure drop.
   - Improver, inadequate or no fan control.

5. **The problem: Not enough supply fan capacity to meet the maximum load.** Undersized supply fan. Peak simultaneous load needed 40,000 CFM, actually available 35,000 CFM.
   - Actual diversity less than design.
   - Simultaneous peak load greater than anticipated. Chose 70 percent fan when it should have been 80 percent.
   - Fan performance data used directly from catalog without correcting rpm and brake horsepower to compensate for inlet vanes.
   - Improper, inadequate or no fan capacity control.

6. **Other fan problems:**
   - Gigantic motors over 125 or 150 hp are cumbersome and a disaster, costwise, to replace.
   - Adjustable drives over 15 hp don't always run smooth. Matching belts and alignment are difficult. On a fix-belt drive four belts may be pulling and two slipping and flopping.
   - Large fans which have to be shipped disassembled, split in the middle, etc. are often out of balance after they are put back together and installed at the job site. They should be field checked and rebalanced if necessary. Splits above the bearing are better.
   - Noise can be a problem, depending on the location of the fan and ductwork in relation to surrounding noises and the extent of system-generated noises.
   - Reverse rotation of a non-operating fan when multiple fans are installed in parallel is a problem when the fan is restarted. Belts can burn up, loud screeching noises develop, starters kick out.
   - Maintenance on direct-drive vaneaxial fans is frequently suspended. Changing a motor involves removing the duct connection, reaching inside the fan, removing the motor enclosure, disconnecting the motor mounting and slipping the fan wheel off the motor shaft.

7. **The problem: Morning warmup or night heating with terminals in minimum open or full shut-off positions.** Various methods or correcting this situation are:
   - Reverse-acting relays in perimeter terminals which switch back after heating is under control. Either reheat coils or central heating equipment are needed in conjunction with the reverse relays.
   - ASHRAE recommends supplying intermittent heat at night, or for an adequate period prior to occupancy, using a central heating coil.
   - If there is a separate perimeter heating system no special provisions need be taken care of by the separate perimeter heating system.

8. **The problem: Startup with unusually high diversity** (say 50 percent) and with normally open terminals. With all the terminals open the system static pressure at startup can be too low to actuate some terminals. When this occurs, terminals must be manually closed until system static pressure is brought up sufficiently to actuate all terminal control.

9. **The problem: On cold winter days, terminals in perimeter zones throttle to minimum settings, or shut off, causing the air in the spaces to become excessively stagnant.** There may also be problems with heating in the perimeter zones, dependent on how the skin heating load is handled with these minimum air conditions.

   One solution to the winter perimeter throttling problem is to add reverse-acting controllers in the perimeter terminals. This allows increased air flow. Reheat coils are needed in this reverse process to avoid cooling the perimeter zones with the increase air flow.

10. **Balancing problems:**
    - Inadequate pressure in end terminal boxes to operate them, or too much air flow through terminal due to excessive pressure at inlet.
    - Diversity between fan CFM and terminals at maximum flow.
    - Pressure imbalance in large systems.
    - Inadequate pressure to operate some boxes at startup because of high CFM flow and low system static pressure.

Chapter 11

ESTIMATING HVAC COSTS

INTRODUCTION

This chapter covers estimating costs for ductwork, piping, equipment, insulation and temperature controls for HVAC and VAV systems in new and existing buildings. It covers estimating retrofit conversions of constant air volume systems to variable air volume systems. Budget and detailed estimating procedures and cost figures are covered. The chapter begins with the general procedure for preparing an estimate and an outline of items required in HVAC and VAV estimates.

THE GENERAL PROCEDURE FOR PREPARING COST ESTIMATES

The general procedure for preparing an HVAC cost estimate is as follows:

- There must be some sort of design completed before any pricing can be done, anything from conceptual to fully detailed. Plus, the type of estimate, whether conceptual, semi-detailed or fully detailed, must correspond to the design and degree of accuracy needed.

- The **quantities and sizes** of the various major HVAC equipment, ductwork, piping, minor equipment and specialties must be listed and material and labor extended and summarized.

- Other peripheral costs for balancing, leak testing, drawings, shipping, etc. must be determined.

- The effect of **existing building** conditions must be considered in remodeling and retrofit projects. Removal of ductwork, piping, equipment, etc. plus cutting and patching must be estimated. See section at end of the chapter.

- Either quotations or budget figures on HVAC **equipment and sub contractors** must be gotten and plugged into the estimate.

- Everything must be **summarized** and added up

- Items such as **taxes,** insurance, rentals of cranes and equipment must be added in.

- The entire estimate be **recapped** with markups for overhead and profit and with potential miscellaneous contingencies .

Opposite is a diagram showing the overall procedure.

ITEMS TO INCLUDE IN AN HVAC ESTIMATE

1. Ductwork such as low, medium and high pressure rectangular galvanized, spiral pipe and fittings, etc.
2. Piping such as black steel, malleable fittings, welded fittings, copper tubing fittings, PVC, etc.
3. Major HVAC equipment such as boilers, chillers, coils, air handling units, fans, etc.
4. Minor HVAC equipment such as grilles, ceiling diffusers, dampers, louvers, piping specialties, valves, etc.
5. Miscellaneous special labor such as balancing, drawings, etc.
6. Cranes, rental equipment, etc.
7. Sub contractors such as insulation and temperature controls.
8. Sales tax, insurance, bonds, etc.
9. Appropriate markups for overhead and profit.

This diagram shows a complete, fast and efficient procedure for preparing piping and sheet metal estimates. The diagram shows the correct sequence of operations and the main areas of work. It follows the critical path method showing the sheet metal estimator, HVAC equipment supplier and subcontractor all preparing their own portions of the estimates at the same time and all coming together for a total bid price within the bid time frame.

Figure 11-1
ESTIMATING PROCEDURE DIAGRAM FOR PIPING AND SHEET METAL WORK

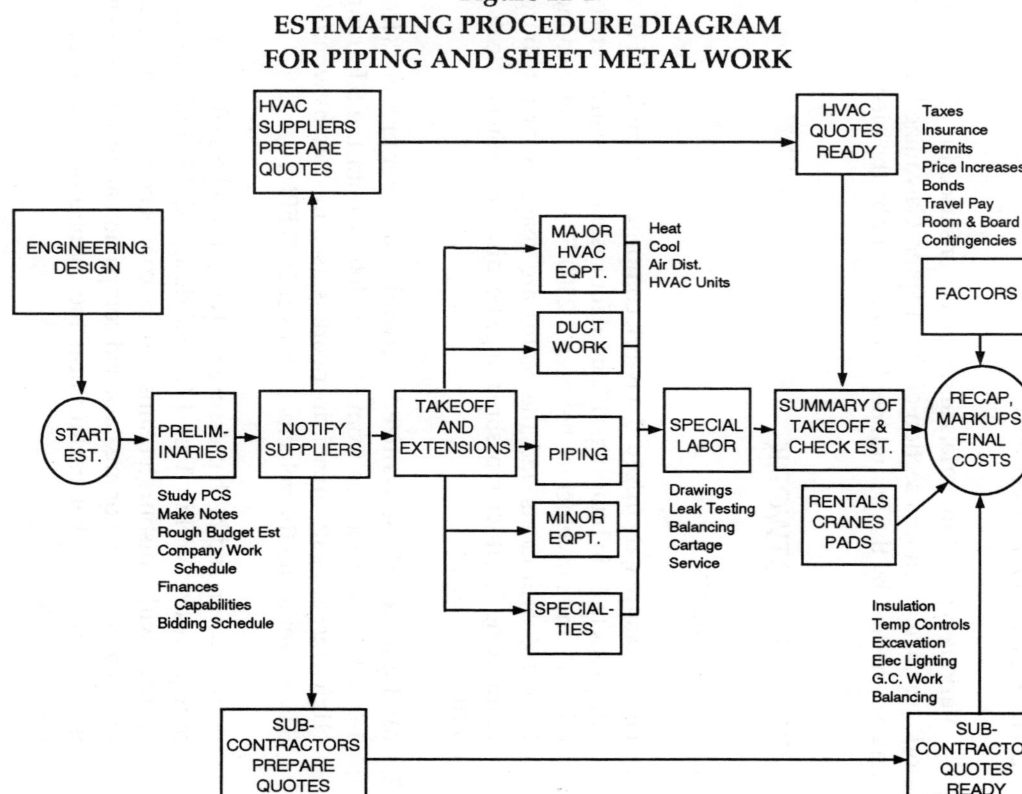

ESTIMATING DUCTWORK

Various types of materials and connections used for HVAC ductwork are covered in this section of the chapter. Galvanized ductwork can be estimated by the pound, per linear foot, or by the piece.

TYPES OF DUCTWORK

HVAC Rectangular

1. **Low pressure galvanized** ductwork comprises the bulk of HVAC ductwork used in buildings. It's used for system pressures between 0-2" S.P. and air velocities between 0-2500 FPM. Generally connections are with cleats and the seams are snaplock or pittsburgh. Reinforcing is either crossbreaking, beading, reinforced cleats or structural angles.

2. **Medium pressure galvanized** ductwork is used for pressures from 2-6" S.P. and velocities from 2000 to 4000 within the S.P. range. **High pressure galvanized** ductwork is used in systems where the S.P. is over 6 and the velocities are over 2000 FPM.

   Both medium and high pressure ductwork must be sealed to maintain pressures within 1 or 1/2% of design CFM. Both are constructed with pittsburgh seams and the connections are with cleats which you can seal,or are gasketed companion angles. Reinforcing is with angles, either backup near the connection and/or at prescribed intervals.

3. **Fiberglass ductboard** is used for about 15% of all the HVAC ductwork in the U.S. It's primarily used for low pressure systems in unconditioned spaces where insulation is needed or for ductwork that requires acoustic insulation. It's easier to fabricate and install than galvanized. Boards are one inch thick, seams and connections shiplap grooved, and stapled and taped.

4. **Aluminum** is used for supply ductwork in HVAC systems if exposed to moisture as in a pool area, shower room etc. It is fabricated the same way as low pressure galvanized with pittsburghs, cleats, angles etc.

HVAC Round

1. Round galvanized spiral **pipe and fittings** are used primarily for high pressure systems with cemented and taped connections but it can also be used for low velocity situations as well as for various types of exhaust systems. The installed cost is slightly less that rectangular galvanized.

2. **Round residential lock seam** ductwork,also known furnace pipe or just galvanized pipe,is most commonly used in residences and apartment buildings as well as for flues. Connections are crimped on one side, slipped together and screwed. Elbows are adjustable.

3. HVAC **flexible tubing,** single skin or factory insulated is used for residential work and for commercial low pressure and high pressure systems.

4. Round **flues,** either single or double skin used for furnaces, unit heaters etc. is the last of the HVAC round ductwork.

TABLE 11-1. BUDGET ESTIMATING GALVANIZED DUCTWORK PER POUND AND PER FOOT

Standard Low Pressure HVAC Rectangular Galvanized 25 Percent Fittings, New Construction, 10 Foot High, 1st Floor

Size	Semi-Perim Inches	Lb/Ft Gauge	Sq Ft/Ft w/20% Waste	No Waste	Selling Price* Furnished & Installed Per Lb.	Per Ft.
6×6	12	26 Ga.	2.8	2.0	$4.79	513.41
12×6	18		3.3.	3.0	4.62	15.24
12×12	24		4.4	4.0	4.44	20.58
18×6	24	24 Ga.	5.6	4.5	4.44	24.91
18×12	30		7.0	5.0	4.31	30.99
24×9	33		7.7	6.5	4.24	32.58
24×12	36		8.4	6.0	4.17	34.96
24×15	39		9.2	6.5	4.17	38.30
30×12	42		9.8	7.0	4.08	40.09
30×18	48		11.2	8.0	4.02	45.00
30×24	54		12.6	9.0	3.95	49.74
36×12	48	22 Ga.	13.6	8.0	3.95	53.68
36×18	54		15.3	9.0	3.85	58.84
36×24	60		17.0	10.0	3.73	63.44
42×12	54		15.3	9.0	3.85	58.84
42×18	60		17.0	10.0	3.73	63.44
42×24	66		18.7	11.0	3.70	69.25
48×12	60		17.0	10.0	3.73	63.14
48×18	66		18.7	11.0	3.70	69.25
48×24	72		20.4	12.0	3.67	74.98
54×24	78		22.1	13.0	3.64	80.56
54×30	84		23.8	14.0	3.62	86.08
54×36	90		25.5	15.0	3.59	91.49

(*Continued*)

TABLE 11-1. BUDGET ESTIMATING GALVANIZED DUCTWORK PER POUND AND PER FOOT (*Continued*)

60×18	78	20 Ga.	26.0	13.0	3.62	94.03
60×24	84		28.0	14.0	7.92	101.26
60×30	90		30.0	15.0	3.59	107.64
72×24	96		32.0	16.0	3.56	113.89
72×30	102		34.0	17.0	3.53	120.05
72×36	108		35.3	18.0	3.49	123.11
84×30	114		38.0	19.0	3.56	130.89
84×36	120		40.0	20.0	3.42	136.63
84×42	126		42.0	21.0	3.38	142.25
96×24	120	18 Ga.	52.6	21.0	3.35	176.64
96×36	132		57.2	22.0	3.33	190.45
96×48	144		62.4	24.0	3.30	205.98
96×72	168		72.8	28.0	3.27	238.24
96×96	192		83.2	32.0	3.24	269.82

*Selling price based on 5.42 per pound for galvanized, $37.00 per hour for labor, a 30 percent markup on direct costs for overhead and a 5 percent markup for profit. The price including material and all labor for the fabrication, installation, drafting and shipping.

Correction factors on selling price of galvanized ductwork.

1. Percent Fittings by Weight, 15% ....0.09 25%.........1.00
   35% ....1.08 45%.........1.15
2. Different wage rates (incl. base pay, fringes, payroll taxes, ins.)
   $12.00 (per hour)...........0.66 $22.00 ....0.89
   15.00...........0.77 26.50 ....1.00
   18.00...........0.80 30.00 ....1.08
3. Overall residential factor for 15 percent fittings, $12.00 per hour gross wages and lighter gauges...0.50 times cost per lb or per foot.

Example: 24"× 8 $4.15/lb equals $2.08/lb
$31-90/ft equals $15.95/ft

Table 11-2
INSTALLED PRICE PER SQUARE FOOT OF DIFFERENT TYPES OF DUCTWORK
Average Size 24"×12" in a Typical Mix
25 Percent Fittings by Square Feet

HVAC Ductwork		Industrial Ductwork	
Spiral	$4.23	Aluminum, Light Ga.	$5.50
Fiber Glass Dct	$4.41	Pvc Coated Galv, Light Ga.	$8.33
Lp Galv, Bare	$5.31	Stainless Steel Light Ga.	$8.83
Lp Galv, Insulated	$7.04	Black Iron, Angle Flanges, 16 Ga.	$11.08
Mp Galv, Bare	$6.53	Black Iron, Angle Flanges, 14 Ga.	$12.32
Mp Galv, Insulated	$8.46	Pvc Plastic	$12.40
		Frp, Fiber Glass Reinforced Plastic	$14.49
		Black Iron, Angle Flanges, 10 Ga.	$16.44

Installed price per square foot includes material, shop labor, field labor, shop drawings, shipping and a 35 percent markup on costs for overhead and profit. Labor is based on $37.00 per hour.

Figure 11-2
Installed Cost per Pound for Galvanized Ductwork

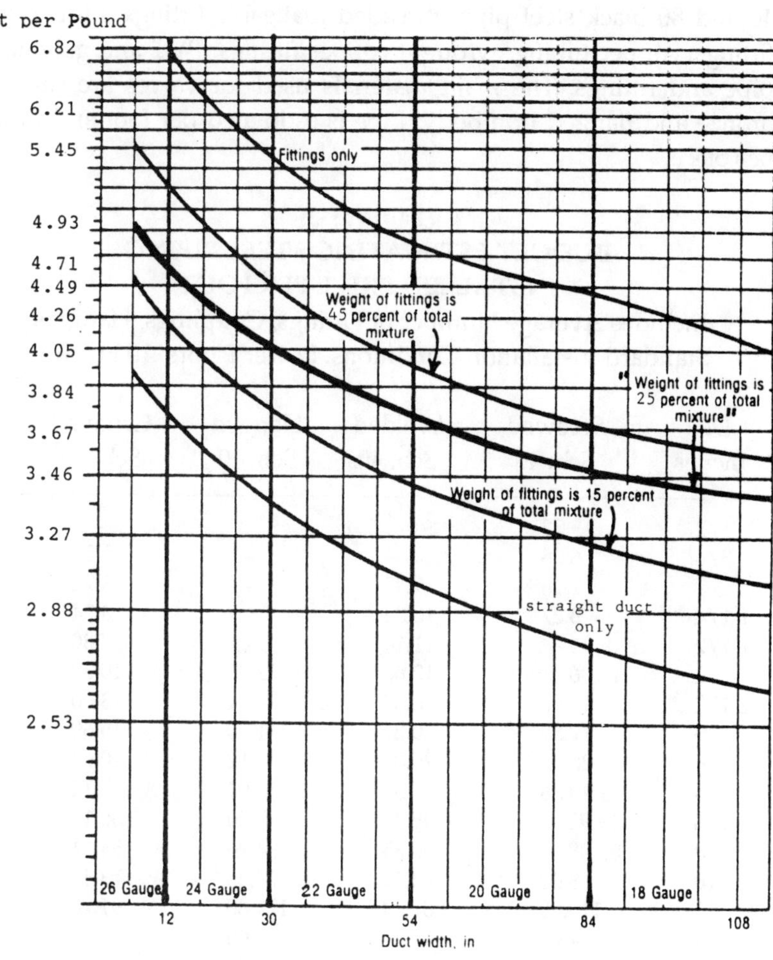

Curves for cost per pound of uninsulated low pressure galvanized ductwork for new construction, 2000 lbs and up, standard installation conditions and conventional duct fabrication (not coil line).

Prices include material, shop and field labor, shop drawings, shipping and a 30 percent markup for overhead and profit on both material and labor.

Costs are based on galvanized material at $.35 per pound and direct labor wage rates as of June 1, 1985 of $27.50 per hour which includes base pay, fringes, insurance and payroll taxes.

ESTIMATING PIPING

The normal piping materials used in HVAC systems are schedule 40 and 80 black steel pipe, threaded malleable fittings, butt welded fittings, copper tubing, wrought copper fittings, PVC and galvanized pipe and fittings. The main materials used for valves are threaded bronze and flanged iron body. Pipe is estimated per foot and fittings per piece.

Table 11-3
BUDGET ESTIMATING STEEL PIPING
INSTALLED PRICE PER FOOT
Includes Average Number of Fittings, Couplings, Hangers, Standard Installation Conditions, Lower Floors, 10 Ft. High

Dia Inches	Screwed Sch. 40	Welded Sch. 40	Flanged Sch. 40	Grooved Sch. 40
1/2	$5.77	$ —	$ _	$—
3/4	6.08	—	—	5.20
1	7.26	9.43	—	5.72
1-1/4	8.27	10.34	—	6.24
1-1/2	9 36	11.14	—	7.80
2	12 06	12.60	22.93	10.30
2-1/2	16.00	16.63	27.14	13.00
3	19.58	20.21	31.20	15.60
4	25.67	26.84	42.12	20.64
5	40.56	37.32	53.04	27.14
6	49 92	45.23	65.00	36.19
8	67 08	56.35	84.76	54.71
10	78.00	64.74	126.36	73.55
12	113.88	87.79	171.60	91.73
14	—	—	199.68	—
16	—	—	230.88	—
18	—	—	249.60	—
20	—	—	283.92	—
22	—	—	293.28	—
24	—	—	309.92	—

Prices include all material costs, labor at $39.00 per hour and overhead and profit markup.

Table 11-4
BUDGET ESTIMATING COPPER TUBING
INSTALLED PRICE PER FOOT
50/50 Solder, 300F
Includes Average Number of Fittings, Couplings, Hangers and Solder.
Standard Installation Conditions, Lower Floors, 10 ft High

DIA Inches	L	M	K	DWV
1/2	$6.30	$5.67	$6.78	5.95
3/4	7.99	7.19	8.94	7.55
1	10.19	8.96	11.29	9.41
1-1/4	12.82	11.54	15.82	12.13
1-1/2	14.66	13.20	16.06	13.86
2	18.92	17.02	20.75	17.88
2-1/2	25.29	22.76	27.70	23.89
3	30.79	27.72	33.72	29.10
4	43.99	39.59	51.32	41.57
5	79.19	71.27	93.67	74.83
6	104.11	93.70	135.14	98.39
8	190.63	154.41	226.93	161.62

Prices include-all material costs, labor at $39.00 per hour and overhead and profit markup.

Correction Factors

1. 95/5 solder 1.07
2. Silver solder 1.16

Table 11-5
THREADED BRONZE AND IRON VALVES

Size	Type of Material	Direct Matl Costs	Labor Man Hrs	Total Matl & Labor Direct Costs	Total Matl & Labor With 30% O & P
Gate					
1/2	Bronze	15.40	0.65	32.95	42.84
3/4	Bronze	19.95	0.72	39.39	51.21
1	Bronze	24.15	0.78	45.21	58.77
1-1/4	Bronze	32.40	0.91	56.97	74.06
1-1/2	Bronze	41.23	0.98	67.69	88.00
2	Bronze	57.43	1.04	85.51	111.16
2-1/2	Iron	191.43	1.80	240.03	312.04
3	Iron	235.60	1.90	286.90	372.97
4	Iron	294.50	2.00	348.50	453.05
5	Iron	0.00	0.00	0.00	0.00
6	Iron	0.00	0.00	0.00	0.00
8	Iron	0.00	0.00	0.00	0.00
Check					
1/2	Bronze	16.50	0.65	34.05	44.27
3/4	Bronze	18.68	0.72	38.12	49.56
1	Bronze	24.45	0.78	45.51	59.16
1-1/4	Bronze	34.50	0.91	59.07	76.79
1-1/2	Bronze	39.00	0.98	65.46	85.10

Table 11-5 (*Continued*)

	2	Bronze	60.00	1.04	88.08	114.50
	2-1/2	Iron	144.00	1.80	192.60	250.38
	3	Iron	172.50	1.90	223.80	290.94
	4	Iron	262.50	2.00	316.50	411.45
	5	Iron	0.00	0.00	0.00	0.00
	6	Iron	0.00	0.00	0.00	0.00
	8	Iron	0.00	0.00	0.00	0.00
Ball						
	1/2	Bronze	5.32	0.65	22.87	29.73
	3/4	Bronze	8.48	0.72	27.92	36.30
	1	Bronze	10.88	0.78	31.94	41.52
	1-1/4	Bronze	18.98	0.91	43.55	56.62
	1/1-2	Bronze	24.15	0.98	50.61	65.79
	2	Bronze	30.00	1.04	58.08	75.50
	2-1/2	Iron	0.00	1.80	48.60	63.18
	3	Iron	0.00	1.90	51.30	66.69
	4	Iron	0.00	2.00	54.00	70.20
	5	Iron	0.00	0.00	0.00	0.00
	6	Iron	0.00	0.00	0.00	0.00
	8	Iron	0.00	0.00	0.00	0.00
Globe						
	1/2	Bronze	21.98	0.65	39.53	51.39
	3/4	Bronze	26.93	0.72	46.37	60.28
	1	Bronze	45.00	0.78	66.06	85.88
	1-1/4	Bronze	64.50	0.91	89.07	115.79
	1-1/2	Bronze	79.50	0.98	105.96	137.75

(*Continued*)

Table 11-5 (*Continued*)

Size	Type of Material	Direct Matl Costs	Labor Man Hrs	Total Matl & Labor Direct Costs	Total Matl & Labor With 30% O & P
2	Bronze	121.50	1.04	149.58	194.45
2-1/2	Iron	367.50	1.80	416.10	540.93
3	Iron	412.50	1.90	463.80	602.94
4	Iron	637.50	2.00	691.50	898.95
5	Iron	0.00	0.00	0.00	0.00
6	Iron	0.00	0.00	0.00	0.00
8	Iron	0.00	0.00	0.00	0.00
Angle					
1/2	Bronze	30.00	0.65	47.55	61.82
3/4	Bronze	39.00	0.72	58.44	75.97
1	Bronze	55.50	0.78	76.56	99.53
1-1/4	Bronze	73.50	0.91	98.07	127.49
1-1/2	Bronze	97.50	0.98	123.96	161.15
2	Bronze	150.00	1.04	178.08	231.50
2-1/2	Iron	0.00	1.80	48.60	63.18
3	Iron	0.00	1.90	51.30	66.69
4	Iron	0.00	2.00	54.00	70.20
5	Iron	0.00	0.00	0.00	0.00
6	Iron	0.00	0.00	0.00	0.00
8	Iron	0.00	0.00	0.00	0.00

Direct labor costs are $37.00 per hour.

ESTIMATING HVAC EQUIPMENT COST AND LABOR

The following major HVAC equipment charts contain the actual cost of the equipment for the different sizes, labor man hours to install, total direct cost and the selling price with the contractor's overhead and profit included. This section is divided into air handling units, fans, VAV equipment, heating equipment and cooling equipment.

Table 11-6
AIR HANDLING UNITS
DX Coil, Electric Heating Coil
Single Zone, Fan and Coil Section
Isolators, Throwaway Filters, Motor, Drives

Tons	Cfm	Heating Mbh	Direct Material Cost			Labor	Total Matl. & Labor	
			Each	Per Ton	Per Cfm	Man Hours	Direct Cost	With 30% O&P
3	1,200	95	$1,863	$621	$1.55	4	$2,011	$2,614
5	2,000	112	2,908	582	1.45	9	3,241	4,213
7.5	3,000	135	4,090	545	1.36	10	4,460	5,798
10	4,000	200	5,204	520	1.30	12	5,648	7,342
12.5	5,000	225	$6,124	$490	$1.22	14	$6,642	$8,635
15	6,000	270	6,832	455	1.14	16	7,424	9,651
20	8,000	360	8,115	406	1.01	18	8,781	11,415
25	10,000	450	8 841	354	0.88	20	9,581	12,455
30	12,000	540	10 333	344	0.86	22	11,147	14,492
40	16,000	675	$12,630	$316	$0.79	26	$13,592	$17,669
50	20,000	810	14,735	295	0.74	30	15,845	20,599
60	24,000	984	17,224	287	0.72	34	18,482	24,026
80	32,000	1,312	22,046	276	0.69	42	23 600	30,680
100	40,000	1,640	26,411	264	0.66	50	28 261	36,739

Air conditioning only, no electric heating coil.
Direct material multiplier .75

Table 11-7
AIR HANDLING UNITS
Single Zone, Fan and Coil Section
Isolators, Throwaway Filters, Motor Drives

Tons	Cfm	Heating Mbh	Direct Material Cost Each	Direct Material Cost Per Ton	Direct Material Cost Per Cfm	Labor Man Hours	Total Matl. & Labor Direct Cost	Total Matl. & Labor With 30% O&P
3	1,200	95	$2 334	$778	$1.94	4	$2,482	$3,226
5	2,000	112	3 641	728	1.82	9	3,974	5,166
7.5	3,000	135	5,121	683	1.71	10	5,491	7,138
10	4,000	200	6,517	652	1.63	12	6,961	9,050
12.5	5,000	225	$7,667	$613	$1.53	14	$8,185	$10,640
15	6,000	270	8,552	570	1.43	16	9,144	11,888
20	8,000	360	10,160	508	1.27	18	10,826	14,074
25	10,000	450	11,070	443	1.11	20	11,810	15,353
30	12,000	540	12,938	431	1.08	22	13,752	17,878
40	16,000	675	$15,813	$395	$0.99	26	$16,775	$21,808
50	20,000	810	18,449	369	0.92	30	19,559	25,427
60	24 000	984	21,564	359	0.90	34	22,822	29,668
80	32 000	1,312	27,602	345	0.86	42	29,156	37,902
100	40,000	1,640	33,066	331	0.83	50	34,916	45,390

Table 11-8
ESTIMATING AIR HANDLING UNITS
2 Row Water or Steam Heating Coil, 6 Row Chilled Water Coil
Single Zone, Fan and Coil Section
Isolators, Throwaway Filters, Motor, Drives

Tons	Cfm	Heating Mbh	Direct Material Cost			Labor	Total Matl. & Labor	
			Each	Per Ton	Per Cfm	Man Hours	Direct Cost	With 30% O&P
3	1,200	95	$1,554	$518	$1.30	4	$1,702	$2,213
5	2,000	112	2,427	485	1.21	9	2,760	3,588
7.5	3,000	135	3,414	455	1.14	10	3,784	4,920
10	4,000	200	4,344	434	1.09	12	4,788	6,225
12.5	5,000	225	$5,111	$409	$1.02	14	$5,629	$7,317
15	6,000	270	5,702	380	0.95	16	6,294	8,182
20	8,000	360	6,771	339	0.85	18	7,437	9,668
25	10,000	450	7,378	295	0.74	20	8,118	10,554
30	12,000	540	8,625	288	0.72	22	9,439	12,271
40	16,000	675	$10,542	$264	$0.66	26	$11,504	$14,955
50	20,000	810	12,300	246	0.61	30	13,410	17,433
60	24,000	984	14,376	240	0.60	34	15,634	20,325
80	32,000	1,312	18,401	230	0.58	42	19,955	25,941
100	40,000	1,640	22,043	220	0.55	50	23,893	31,061

Correction Factors on all Air Handling Units		Material	Labor
1. Multi-zone unit instead of single zone,			
DX and Electric		1.40	1.25
DX and Gas		1.32	1.25
HW and CHW Coils		1.50	1.25
2. Suspended installation instead of floor mounted		1.25	
3. With filter, mixing box (add)		0.29 /CFM	1.05
4. Variable Air Volume (add)		0.51 /CFM	1.10
5. With economizer section (add)	10 tons	$998	1.05
	20 tons	1,229	1.05
	50 tons	1,843	1.05

Table 11-9 — ESTIMATING ROOF TOP UNITS

Single Zone, DX Cooling Electric Heating Coils

Includes Economizers. Coils, Filters, Curbs, Standard Controls, Warranty

Tons	Cfm	Electric Coil	Direct Material Cost			Labor	Total Material & Labor	
		kW	Each	Per Ton	Per Cfm	Man Hours	Direct Cost	With 30% O&P
2	800	15	$1,912	$956	$2.39	4	$2,060	$2,678
3	1,200	20	2,263	754	1.89	4	2,411	3,135
5	2,000	25	3,594	719	1.80	6	3,816	4,960
7.5	3,000	37	6,405	854	2.14	8	6,701	8,712
10	4,000	50	8,410	841	2.10	8	8,706	11,318
12.5	5,000	62	$10,422	$834	$2.08	10	$10,792	$14,030
15	6,000	75	12,314	821	2.05	10	12,684	16,489
20	8,000	87	16,102	805	2.01	12	16,546	21,509
25	10,000	100	19,767	791	1.98	12	20,211	26,274
30	12,000	125	$23,292	$776	$1.94	16	$23,884	$31,049
40	16,000	150	30,478	762	1.90	20	31,218	40,583
50	20,000	190	37,378	748	1.87	22	38,192	49,650
60	24,000	220	43,129	719	1.80	26	44,091	57,319

CORRECTION FACTORS		Material	Labor
1. Variable air volume unit		1.28	1.10
2. Multizone units		1.33	1.25
3. Heat Pumps (with electric heating coils)		1.30	
4. Cooling only, no heating		0.66	
5. Power return fan section, add direct costs	10 tons	$1,629	1.10
	20 tons	2,442	1.10
	50 tons	5,373	1.10
6. Omit economizer, deduct	10 tons	$1,058	0.95
	20 tons	1,303	0.95
	50 tons	1,953	0.95

Table 11-10 — ESTIMATING ROOF TOP UNITS

Single Zone, DX Cooling, Gas Heating, Staged Cooling and Heating, 7-1/2 Ton and Up
Includes Economizers, Coils, Curbs, Filters, Standard Controls, Warranty

Tons	Cfm	Heating Mbh	Direct Material Cost			Labor	Total Material & Labor	
			Each	Per Ton	Per CFM	Man Hours	Direct Cost	With 30% O&P
2	800	60	$2,185	$1,092	$2.73	6	$2,407	$3,129
3	1,200	94	2,516	839	2.10	8	2,812	3,655
5	2,000	112	3,993	799	2.00	10	4,363	5,672
7.5	3,000	135	7,127	950	2.38	12	7,571	9,843
10	4,000	200	9,345	934	2.34	12	9,789	12,725
12.5	5,000	225	$11,581	$926	$2.32	14	$12,099	$15,728
15	6,000	270	13,680	912	2.28	14	14,198	18,458
20	8,000	360	17,890	894	2.24	16	18,482	24,026
25	10,000	450	21,960	878	2.20	16	22,552	29,318
30	12,000	540	$25,877	$863	$2.16	18	$26,543	$34,505
40	16,000	675	33,864	847	2.12	24	34,752	45,177
50	20,000	810	41,532	831	2.08	26	42,494	55,242
60	24,000	985	48,879	815	2.04	32	50,063	65,081

CORRECTION FACTORS		Material	Labor
1. Variable air volume unit		1.25	1.10
2. Multizone units		1.30	1.25
3. Hot water heating coils		0.90	—
4. Steam heating coil		0.93	—
5. Power return fan section, add direct costs	10 tons	$1,536	1.10
	20 tons	2,304	1.10
	50 tons	5,069	1.10
6. Omit economizer, deduct	10 tons	$998	0.95
	20 tons	1,229	0.95
	50 tons	1,843	0.95

Table 11-11
SELF-CONTAINED AIR CONDITIONING UNITS
Air Cooled, DX Coil, Condenser Section,
Electric Heating Coil, Supply Fan, Filters

Tons	Cfm	Heating Mbh	Direct Material Cost			Labor	Total Matl. & Labor	
			Each	Per Ton	Per Cfm	Man Hours	Direct Cost	With 30% O&P
3	1,200	94	$2,995	$998	$2.50	6	$3,217	$4,182
5	2,000	115	3,442	688	1.72	12	3,886	5,051
7.5	3,000	135	4,934	658	1.64	14	5,452	7,088
10	4,000	200	6,325	632	1.58	15	6,880	8,944
12.5	5,000	225	7,787	623	1.56	18	8,453	10,989
15	6,000	270	9,201	613	1.53	20	9,941	12,923
20	8,000	360	12,043	602	1.51	22	12,857	16 714
25	10,000	450	14,177	567	1.42	24	15,065	19 584
30	12,000	540	16,292	543	1.36	26	17,254	22,431
40	16,000	675	20,126	503	1.26	31	21,273	27,655
50	20,000	810	23,401	468	1.17	36	24,733	32,153
60	24,000	985	27,794	463	1.16	42	29,348	38,153

CORRECTION FACTORS	Material	Labor
1. Water cooled		
2. Water or steam heating coils.		

Direct labor costs are $37.00 per hour.

Table 11-12
ESTIMATING CENTRIFUGAL FANS
Air Foil Wheel, Single Width Single Inlet, Outlet Velocity 2000 Fpm
Motors, Drives, Isolators Included

Total Costs			Inches	Material Cost		Labor To	Total Matl & Labor	
Cfm	Hp	Wheel Diam.	Static Press.	Total	Per Cfm	Install Man Hrs	Direct Costs	Sell With 30% O&P
1,000	1/3	12	1"	$ 508.56	$.51	3	$ 589.68	$ 766.48
2,000	1/2	15	1"	917.28	.47	7	1,103.44	1,434.16
4,000	2	18	1"	1,220.96	.30	10	1,489.28	1,935.44
6,000	2-1/2	22	1-1/2"	1,526.72	.26	12	1,848.08	2,404.48
8,000	3	27	1-1/2"	1,934.40	.25	14	2,308.80	3,002.48
10,000	5	30	2"	2,291.12	.24	16	2,719.60	3,534.96
12,000	10	33	2"	2,694.64	.24	18	3,175.12	4,127.76
14,000	10	33	2"	3,207.36	.24	20	3,742.96	4,865.12
16,000	10	36-1/2	2"	3,463.20	.23	22	4,050.80	5,267.60
18, 000	10	40-1/4	2"	3, 895.84	.23	24	4, 536.48	5,898.88
20,000	10	44-1/4	2"	4,073.68	.21	26	4,769.44	6,199.44
25,000	15	44-1/4	2-1/2"	5,092.88	.21	30	5,894.72	7,662.72
30,000	20	54-1/4	2-1/2"	6,111.04	.21	36	7,074.08	9,196.72
40,000	25	60	2-1/2"	7,638.80	.20	46	8,870.16	11,531.52
50,000	30	66	3"	9,549.28	.20	56	11,046.88	14,360.32
60,000	40	73	3"	11,458.72	.20	64	13,170.56	17,122.56

Table 11-13
ESTIMATING UTILITY SETS
Forward Curve Wheels, Single Width Single Inlet
Motors, Drives, Isolators Included

Cfm	Hp	Wheel Diam.	Inches Static Press.	Material Cost Each	Material Cost Per Cfm	Labor To Install Man Hr.	Total Matl & Labor Direct Costs	Total Matl & Labor Sell With 30% O&P
500	1/4	10	1/2"	$ 253.76	$.51	2	$ 302.64	$ 401.44
1,000	1/2	12	3/4 "	431.60	.44	4	538.72	700.96
2,000	1	12	1"	712.40	.36	5	850.72	1,101.36
4,000	2	18	1"	814.32	.21	7	1,001.52	1,302.08
6,000	3	24	1-1/4"	1,145.04	.20	9	1,386.32	1,802.32
8,000	5	27	1-1/"	1,323.92	.17	10	1,591.20	2,068.56
10,000	5	30	1-1/2"	1,526.72	.16	12	1,848.08	2.404.48
12,000	5	33	1-1/2"	1,832.48	.16	14	2,206.88	2,869.36
14,000	7-1/2	36	1-1/2 "	2,138.24	.16	16	2,566.72	3,336.32
16,000	7-1/2	39	1-1/"	2,240.16	.15	18	2,721.68	3,538.08
18,000	10	40	1-1/2"	2,520.96	.15	20	3,054.48	3,971.76
20,000	15	44	1-1/2"	2,800.72	.15	22	3,389.36	4,405.44

Table 11-14
ESTIMATING ROOF EXHAUST FANS
Centrifugal, Belt Driven, Aluminum Housing, 1/2" Static Pressure
with Shutter, Birdscreen, Curb

			Material Cost		Labor To	Total Matl & Labor	
Cfm	Hp	Wheel Diam.	Static Each	Per Per Cfm	Install Man Hrs	Direct Costs	Sell With 30% O&P
500	1/12	10	$ 449.28	$.90	2	$ 504.40	$ 656.24
1,000	1/6	12	535.60	.53	3	556.40	723.84
1,500	1/4	14	538.72	.42	3	685.36	889.20
2,000	1/3	22	662.48	.34	4	770.64	1,001.52
3,000	1/2	24	749.84	.26	4	855.92	1,114.88
4,000	3/4	24	865.28	.23	5	999.44	1,298.96
6,000	1	30	1,030.64	.18	6	1,191.84	1,548.56
8,000	1-1/2	30	1.220.96	.16	6	1,275.04	1,657.76
10,000	2	36	1,488.24	.16	7	1,675.44	2,177.76
15,000	3	48	2,234.96	.16	8	2,448.16	3,182.40
20,000	5	48	3,077.36	.16	9	3,317.60	4,313.92

Table 11-15
ESTIMATING VANE-AXIAL FANS
Automatic Controllable Pitch, Direct Drive Includes Inlet Cones, T Frame Motors, Horizontal Supports, Isolators, Pneumatic Actuator

Total Costs			Inches	Material Cost		Labor To	Total Matl & Labor	
Cfm	Hp	Wheel Diam.	Static Press.	Total	Per Cfm	Install Man Hrs	Direct Costs	Sell With 30% O&P
Supply Fans Medium Pressure, 1770 RPM								
20,000	30	36/26	5"	$9,115.60	.46	22	$9,726.08	$12,643.28
40,000	50	42/26	5"	10,201.36	.26	32	11,089.52	14,416.48
60,000	100	48/26	6"	12,670.32	.21	44	13,892.32	18,059.60
80,000	150	48/26	6"	14,430.48	.20	48	16,763.76	21,793.20
110,000	200	54/26	6"	17,914.00	.17	64	19,691.36	25,599.60
Return Air Fans, 1170 RPM								
18,000	10	38/26	2"	8,589.36	.48	20	9,144.72	11,888.24
36,000	20	45/26	2"	9,502.48	.27	36	10,500.88	13,653.12
54,000	30	54/26	2"	10,686.00	.20	48	12 019.28	15.624.96
72,000	50	54/26	2"	11,663.60	.17	50	13,052.00	16,771.04
100,000	75	60/26	2"	14,954.16	.16	64	16,732.56	21,753.68

Above fan selection and prices based on Joy vane-axial fans
Direct Labor costs are $37.00 per hour.

Table 11-16
INLET VANE DAMPERS
For Centrifugal Fans

Dia. Inches	Area Sq ft	Direct Material Costs Each	Direct Material Costs Per Sq ft	Labor Man Hrs	Total Matl & Labor Direct Costs	Total Matl & Labor With 30% O&P
18	1.80	70.20	39.00	1.50	109.20	141.96
20	2.20	77.00	35.00	1.70	121.20	157.56
22	2.60	85.28	32.80	1.90	134.68	175.08
24	3.10	94.55	30.50	2.00	146.55	190.52
26	3.70	103.60	28.00	2.20	160.80	209.04
28	4.30	117.18	27.25	2.30	176.98	230.07
30	5.00	128.63	25.73	2.40	191.03	248.34
33	6.00	144.55	24.09	2.60	212.15	275.80
36	7.10	163.30	23.00	2.80	236.10	306.93
39	8.30	182.60	22.00	3.10	263.20	342.16
42	9.60	201.60	21.00	3.30	287.40	373.62
45	11.00	220.00	20.00	3.50	311.00	404.30
48	12.60	239.40	19.00	4.00	343.40	446.42
54	16.00	286.20	17.89	4.60	405.80	527.54
60	19.60	348.69	17.79	5.30	486.49	632.44
66	23.80	411.74	17.30	6.00	567.74	738.06
72	28.20	481.10	17.06	6.50	650.10	845.13
78	33.20	561.74	16.92	7.10	746.34	970.24
84	38.50	650.65	16.90	7.60	848.25	1102.73
90	44.10	742.56	16.84	7.80	945.36	1228.97
96	50.30	840.01	16.70	8.00	1048.01	1362.41

Direct labor costs are $37.00 per hr.

Table 11-17
ESTIMATING REGISTERS
Return Air Registers, Fixed 45° Vanes, Opposed Blade Dampers, Commercial Grade

Size	Semi-Perim Inches	Sq Ft	Direct Material Cost Each	Direct Material Cost Per Sq Ft	Labor Man Hours	Total Matl. & Labor Direct Cost	Total Matl. & Labor With 30% O&P
12×6	18	0.5	$28.75	$57.49	0.8	$58	$76
12×12	24	1.0	37.55	37.55	0.9	71	92
18×12	30	1.5	50.96	33.98	1.0	88	114
18×18	36	2.3	69.83	30.36	1.1	111	144
24×12	36	2.0	60.69	30.35	1.1	101	132
24×24	48	4.0	115.02	28.76	1.3	163	212
30×12	42	2.5	$76.87	$30.75	1.2	$121	$158
30×18	48	3.8	112.91	29.71	1.3	161	209
30×24	54	5.0	139.78	27.96	1.5	195	254
36×18	54	4.5	127.65	28.37	1.5	183	238
36×36	72	9.0	233.37	25.93	1.7	296	385
42×24	66	7.0	188.98	27.00	1.6	248	323
48×24	72	8.0	210.86	26.36	1.7	274	356
48×48	96	16 0	452.06	28 25	2.6	548	713
54×18	72	6.8	$183.71	$27.02	1.7	$247	$321
54×24	78	9.0	234.35	26.04	2.0	308	401
60×24	84	10.0	270.98	27.10	2.3	356	463
60×48	108	20.0	495.19	24.76	3.0	606	788

Correction Factors-Commercial Grades	Material	Labor
1. Supply registers, single deflection dampers	1.10	1.00
2. Supply registers, double deflection dampers	1.25	1.00
3. Transfer grille, single deflection, no dampers	0.63	0.60
4. Relief grille in ceiling, lay in, single deflection, no dampers	0.63	0.40
5. Aluminum construction instead of steel	1.10	1.00
6. Lay in type grilles	1.00	0.50
7. Side wall screw in	1.00	0.65
8. Residential light commercial grade	0.60	0.95

Table 11-18
CEILING DIFFUSERS
Round, Fixed Pattern, Steel, w/OBD. Commercial Grade

Neck Diameter	Direct Material Cost		Labor	Total Material & Labor	
Inches	Each	Per In. of Dia.	Man Hours	Direct Cost	With 30% O&P
6	$43.70	$7.28	0.7	$69.60	$90.49
8	55.75	6.97	0.8	85.35	110.96
10	66.31	6.63	0.9	99.61	129.49
12	76.85	6.40	1.0	113.85	148.01
14	102.48	7.32	1.1	143.18	186.13
16	131.10	8.19	1.2	175.50	228.15
18	159.74	8.87	1.3	207.84	270.19
20	183.84	9.19	1.5	239.34	311.15
24	220.62	9.19	1.6	279.82	363.77
30	275.78	9.19	1.9	346.08	449.90
36	330.92	9.19	2.2	412.32	536.02
Round, Adjustable Pattern, Steel, w/OBD, Commercial Grade					
6	$51.24	$8.54	0.7	$77.14	$100.28
8	64.80	8.10	0.8	94.40	122.73
10	76.85	7.69	0.9	110.15	143.20
12	88.91	7.41	1.0	125.91	163.69
14	119.05	8.50	1.1	159.75	207.68
16	152.20	9.51	1.2	196.60	255.58
18	185.36	10.30	1.3	233.46	303.50
20	213.98	10.70	1.5	269.48	350.33
24	256.79	10.70	1.6	315.99	410.79
30	320.99	10.70	1.9	391.29	508.67
36	385.78	10.72	2.2	467.18	607.33

(Continued)

Table 11-18 (*Continued*)

Rectangular, Adjustable Pattern, Steel, w/OBD, Commercial Grade

6×6	$31.95	—	0.7	$57.85	$75.20
9×9	38.87	—	0.9	72.17	93.83
12×12	48.52	—	1.1	89.22	115.99
15×15	41.15	—	1.2	85.55	111.22
18×18	102.77	—	1.5	158.27	205.75
21×21	124.93	—	1.6	184.13	239.37
24×24	145.42	—	1.7	208.32	270.82

CORRECTION FACTORS	MATERIAL	LABOR
1. Aluminum	1.10	1.00
2. Lay-in diffusers	0.65	0.60

Table 11-19

VAV TERMINAL BOXES

Cooling Only With Pneumatic or Electric Motor, Controls

Cfm Range	Coil Sq Ft	Direct Material Cost	Labor	Material & Labor	
			Man Hours	Direct Cost	With 30% O&P
200-400	—	$312.87	2.0	$387	$503
400-600	—	368.85	2.7	469	609
600-800	—	429.12	3.3	551	717
800-1,000	—	476.49	4.0	624	812
1,000-1,500	—	531.02	4.5	698	907
1,500-2,000	—	595.61	5.0	781	1,015
2,000-3,000	—	645.84	5.5	849	1,104

Table 11-20
VAV TERMINAL BOXES
With Reheat Coils, Pneumatic or Electric Motor and Controls

Cfm Range	Coil Sq Ft	Direct Material Cost	Labor	Material & Labor	
			Man Hours	Direct Cost	With 30% O&P
200-400	0.8	$452.09	2.7	$552	$718
400-600	1.0	519.54	2.9	627	815
600-800	1.5	625.75	3.6	759	987
800-1,000	2.0	687.46	4.3	847	1,101
1,000-1,500	2.5	851.07	4.9	1,032	1,342
1,500-2,000	3.0	937.19	5.5	1,141	1,483
2,000-3,000	3.5	1016.12	6.0	1,238	1,610

Figure 11-3
VAV BOX RETROFIT KITS

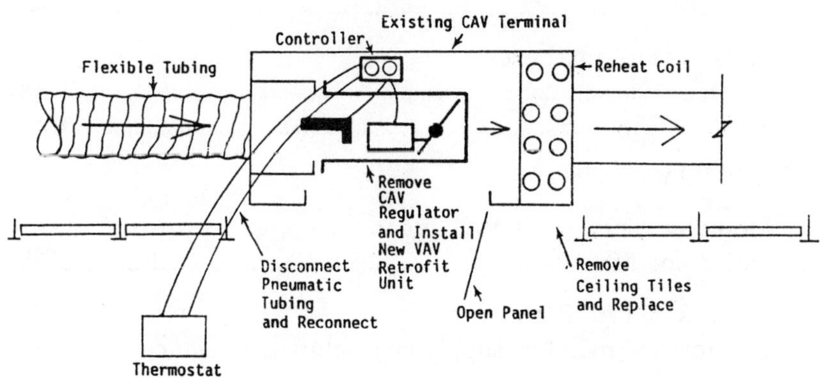

Cost Each

1. For Buensod
   Up to 1000 Cfm $360
2. Titus kits for Titus boxes
   0-500 .... $260
   500-1000 Cfm $360

Figure 11-4. VAV FLOW CONTROLLER
Add to Existing CAV Box Intake and Disengage CAV on Box

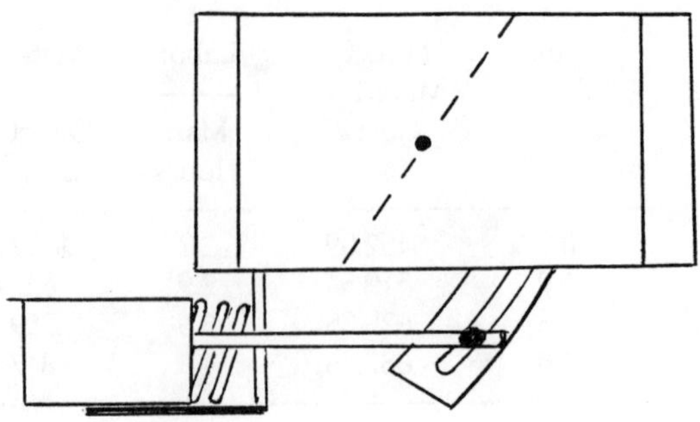

VAV SYSTEMS COMPONENTS

TITUS VAV BOX RETROFIT KITS

Cost Each

1. For Buensod or Tutle and Baily CAV boxes
   Up to 1000 Cfm $345

2. Titus kits for Titus boxes
   0-500 .... $248
   500-1000 Cfm $345

CARRIER VAV UNITS

1. Moduline cooling only VAV boxes
   with 3 slot diffusers $235

VAV CONTROL PANELS

1. Controls and monitors supply fans, return fans, static pressure sensor, supply and return air, air monitoring stations, transmitters, outside air control .... $8,280 to $13,800

2. Static pressure regulator $207

SLOT COOLING ONLY VAV BOX

1. Moduline cooling only VAV boxes
   with 3 slot diffusers $240

Table 11-21
VAV CEILING DIFFUSERS

Dia. Inches	Cfm Range	Direct Costs Each	Labor Man hrs	Total Matl & Labor Direct Costs	With 30% O&P
6	100-220	$240	2	$360	$468
8	160-355	$240	2	$360	$468
10	260-580	$240	2	$360	$468
12	380-890	$240	2	$360	$468

Self-contained variable volume ceiling diffusers with thermal sensors for discharge air and warmup, built-in volume controls and built-in pressure sensor for automatic switch-over between cooling and heating. Direct labor costs are $40 per hour.

Table 11-22
Air Flow Measuring Stations

Size	Sq Ft	Direct Material Costs Each	Per Sq Ft	Labor Man Hr	Total Matl & Labor Direct Costs	With 30% O&P
18x12	1.50	108.00	72.00	1.30	143.10	186.03
24x12	2.00	130.00	65.00	1.40	167.80	218.14
24x24	4.00	225.00	56.25	1.60	268.20	348.66
36x24	6.00	300.00	50.00	2.20	359.40	467.22
36x36	9.00	405.00	45.00	2.40	469.80	610.74
48x24	8.00	375.00	46.88	2.60	445.20	578.76
48x36	12.00	480.00	40.00	3.00	561.00	729.30
60x24	10.00	430.00	43.00	3.20	516.40	671.32
60x32	15.00	585.00	39.00	3.50	679.50	883.35
72x36	18.00	684.00	38.00	3.80	786.60	1022.58
72x42	21.00	777.00	37.00	4.30	893.10	1161.03

Direct labor costs are $37.00 per hour.

1. Controls and monitors supply fans, return fans, static pressure sensor, supply and return air, air monitoring stations, transmitters, outside air control .... $6000 to $10,000
2. Static pressure regulator $150

Table 11-23
GAS FIRED CAST IRON BOILERS
Hot Water and Steam

Heating Btuh	Cfm 70		Direct Material Cost	Labor		Total Material & Labor	
Input	Output	Rise	Each	Per 1,000 Btu	Man Hours	Direct Cost	With 30% O&P
146,000	127,000	110,000	$1,695	$11.61	16	$2,287	$2,973
251,000	218,000	188,000	2,411	9.61	23	3,262	4,240
352,000	306,000	264,000	3,164	8.99	26	4,126	5,364
450,000	391,000	337,000	3,654	8.12	29	4,727	6,145
548,000	477,000	411,000	4,896	8.93	31	6,043	7,856
830,000	722,000	622,000	7,119	8.58	37	8,488	11,035
1,012,000	880,000	759,000	7,535	7.45	46	9,237	12,008
1,394,000	1,212,000	1,056,000	10,398	7.46	50	12,248	15,923
1,624,000	1,412,000	1,248,000	12,206	7.52	54	14,204	18,465
1,980,000	1,722,000	1,537,000	14,617	7.38	65	17,022	22,129

Table 11-23 (*Continued*)

2,320,000	2,017,000	1,801,000	17,180	7.41	74	19,918	25,893
2,784,000	2,421,000	2,162,000	20,194	7 25	82	23,228	30,196
3,092,000	2,689,000	2,400,000	22,152	7.16	88	25,408	33,030
3,710,000	3,226,000	2,880,000	26,071	7.03	95	29,586	38,462
4,330,000	3,765,000	3,361,000	30,592	7.07	112	34,736	45,157
4,950,000	4,304,000	3,843,000	34,057	6.88	122	38,571	50,142
6,180,000	5,374,000	4,800,000	51,237	8.29	144	56,565	73,534

1. Includes valves, controls, insulated jacket.
2. Oil fired hot water or steam versus gas fired, .96 multiplier on material.
3. Add for air eliminator package which includes expansion tank, air vent and fill valve.

110 - 264 MBH	$100.96
337 - 759 MBH	$111.52
1,056 - MBH & UP	$129.60

4. Gross output is the amount of heat needed to heat the building plus piping radiation losses in the distribution system plus pickup allowances for warmups, etc.
5. Net output is the amount needed to heat the building.

Direct labor costs are $37.00 per hour.

Table 11-24
BASEBOARD HEATING
Per Foot

Description	Size	Btuh	Material Cost Per Ft	Labor Man Hours	Total Material & Labor per Ft With Direct	Total Material & Labor per Ft 30% O&P
HOT WATER RADIATION (Panel, Cast Iron, with Damper, Fin Tube, Wall Hung, Supports, Excluding Covers)			$23.81	0.8	$53.41	$69.43
ALUMINUM FIN Copper Tube	1-1/4 "		30.98	1	67.98	88.38

Table 11-24 (*Continued*)

STEEL FIN					
Steel Tube	1-1/4 "	27.32	1	64.32	83.62
Steel Tube	2 "	31.05	1.3	79.15	102.90
PACKAGE					
ALUMINUM FIN					
Copper Tube	1/2 "	7.59	0.62	30.53	39.69
Copper Tube	3/4 "	7.94	0.67	32.73	42.54
Copper Tube	1 "	13.80	0.67	38.59	50.17
Copper Tube	1-1/4 "	19.67	0.67	44.46	57.79
STEEL FIN					
Iron Pipe Size					
Steel Tube		22.08	0.67	46.87	60.93
CONVECTOR UNIT					
(Damper,Flush,Trim,					
Floor Indented)		36.92	0.8	66.52	86.47

Table 11-25
HOT WATER REHEAT COILS
Flanged, 2 Row—Water 200°F in, 180°F out, 3-4 FPS—Air 70°F in, 110°F out, 700 FPM

Btuh	Typical Size	Sq Ft	GPM	CFM	Direct Material Cost		Labor	Total Material & Labor	
		Area			Each	Per Sq Ft	Man Hours	Direct Costs	With 30% O&P
15,100	12×6	0.5	1.5	350	$188	$375.36	1.7	$251	$326
30,200	12×12	1.0	3.0	700	218	218.04	1.8	285	370
45,400	18×12	1.5	4.5	1,050	246	163.76	2.0	320	416
60,500	24×12	2.0	6.1	1,400	270	135.24	2.2	352	457
75,600	24×18	2.5	7.6	1,750	297	118.68	2.4	386	501
90,700	36×12	3.0	9.1	2,100	317	105.80	2.6	414	538
105,800	36×14	3.5	10.6	2,450	345	98.57	2.8	449	583
121,000	36×16	4.0	12.1	2,800	373	93.15	3.0	484	629
151,200	36×20	5.0	15.1	3,500	455	91.08	3.5	585	760
181,400	48×18	6.0	18.1	4,200	552	92.00	4.0	700	910
241,800	48×24	8.0	24.1	5,600	718	89.70	5.0	903	1173

Includes complete installation of coil in duct. Does not include valves or piping connections.
Correction Factors
1. 6 Row Coils, Material add 25%, Labor add 15%.
Direct labor costs are $37.00 per hour.

Table 11-26
RECIPROCATING CHILLERS
Water Cooled With Multiple Semi-Hermatic Compressors

Tons	Direct Material Cost		Labor	Total Material & Labor	
	Each	Per-Ton	Man Hours	Direct Cost	With 30% O&P
20	$14,166	$708	28	$15,202	$19,762
40	21,097	527	32	22,281	28,966
60	28,180	470	36	29,512	38,365
80	35,866	448	40	37,346	48,550
100	43,702	437	48	45,478	59,121
120	51,840	432	56	53,912	70,085
140	57,717	412	62	60,011	78,014
160	64,196	401	66	66,638	86,630
180	69,552	386	70	72,142	93,785

Table 11-27
RECIPROCATING CHILLERS
Air Cooled With Compressor

Tons	Each	Per-Ton	Man Hours	Direct Cost	With 30% O&P
20	$15,823	$791	28	$16,859	$21,917
40	23,358	584	32	24,542	31,904
60	31,568	526	36	32,900	42,769
80	40,688	509	40	42,168	54,818
100	47,258	473	48	49,034	63,744
120	59,299	494	56	61,371	79,782

Correction Factors	Materials	Labor
1. Air cooled reciprocating chiller without condenser	0.72	0.75

Labor includes unloading from truck at job site, staging, uncrating, hoisting and setting into place, anchoring, aligning, starting and checkout.

Manhours do not include piping, valves or electrical hookup Based on direct labor costs of $37.00

Table 11-28
CENTRIFUGAL WATER COOLED CHILLERS
With Condenser and Single Compressor

Tons	Approx Weight	Direct Material Cost		Labor	Total Material & Labor	
		Each	Per Ton	Man Hours	Direct Cost	With 30% O&P
100	7,500	$64,046	$640	76	$66,858	$86,915
200	10,000	72,334	362	80	75,294	97,882
300	12,000	85,897	286	88	89,153	115,899
400	18,000	102,473	256	92	105,877	137,640
500	20,000	119,050	238	95	122,565	159,334
600	24,000	137,133	229	98	140,759	182,987
700	24,000	158,231	226	102	162,005	210,606
800	27,000	176,314	220	106	180,236	234,307
900	33,000	195,905	218	112	200,049	260,063
1,000	37,000	213,988	214	118	218,354	283,861
1,200	43,000	253,169	211	124	257,757	335,084
1,400		289,336	207	132	294,220	382,486
1,600		327,010	204	140	332,190	431,847

Labor includes receiving package chiller at job site, unloading, uncrating, setting in place, aligning, etc.
Labor does not include piping, valves or electrical hookup.
Crane rental costs not included.
Chillers with DOUBLE BUNDLE condensers run $80.00 more per ton.
Direct labor costs are $37.00 per hour.

Table 11-29
COOLING TOWERS

Tons	Gpm	Fan Hp	Direct Material Costs Each	Direct Material Costs Per Ton	Labor Man Hours	Total Material & Labor Direct Costs	Total Material & Labor With 30% O&P
Crossflow, induced draft, propeller fan, shipped unassembled*							
100	240	5	$5,658	$57	130	$10,468	$13,608
200	480	7.5	10,764	54	190	17,794	23,132
300	720	15	13,593	45	260	23,213	30,177
400	960	20	15,594	39	330	27,804	36,145
500	1,200	20	18,009	36	370	31,699	41,209
600	1,440	25	20,424	34	400	35,224	45,791
800	1,920	30	25,530	32	450	42,180	54,834
1,000	2,400	35	28,980	29	550	49,330	64,129
Counter flow, forced draft, centrifugal fan, shipped unassembled*							
100	240		$7,728	$77	137	$12,797	$16,636
200	480		14,766	74	200	22,166	28,816
300	720		17,250	58	273	27,351	35,556
400	960		22,770	57	347	35,609	46,292
500	1,200		27,738	55	389	42,131	54,770
600	1,440		32,706	55	420	48,246	62,720
800	1,920		43,470	54	473	60,971	79,262
1,000	2,400		54,510	55	578	75,896	98,665
Induced draft, propeller, galvanized, shipped assembled							
10	24	750 lb	$1,725	$173	12	$2,169	$2,820
15	36	900	2,484	166	13	2,965	3,855
20	48	980	3,174	159	14	3,692	4,800
25	60	1,000	3,795	152	16	4,387	5,703
30	72	1,180	4,347	145	17	4,976	6,469
40	96	1,290	5,520	138	18	6,186	8,042
50	120	1,800	6,072	121	20	6,812	8,856
60	144	2,300	6,293	105	22	7,107	9,239
80	192	3,500	7,066	88	24	7,954	10,340

(Continued)

Table 11-29
Cooling Towers (*Continued*)

Tons	Gpm	Fan Hp	Direct Material Costs		Labor	Total Material & Labor	
			Each	Per Ton	Man Hours	Direct Costs	With 30% O&P
100	240	4,300	8,280	83	27	9,279	12,063
125	300	4,900	10,005	80	29	11,078	14,401
150	360	6,000	11,592	77	32	12,776	16,609
175	420	6,200	13,041	75	34	14,299	18,589
200	480	6,900	14,352	72	36	15,684	20,389
300	720	9,000	20,700	69	51	22,587	29,363
400	960	14,500	26,496	66	63	28,827	37,475
500	1,200	16,900	31,740	63	72	34,404	44,725

*Redwood, Treated Fir

Man hours include unloading, handling, assembling and set in place. Piping, electrical wiring or crane rental are not included.

Direct labor costs are $37.00 per hour.

Table 11-30
HEAT PUMPS

Tons of Cooling	Heating Capacity	Direct Material Costs		Labor	Total Material & Labor	
		Each	Per Ton	Man Hours	Direct Costs	With 30% O&P
Split System, Air to Air						
2	8.5	$1,573	$570	9	$1,906	$2,478
3	13	2,167	523	10	2,537	3,298
5	27	3,640	528	20	4,380	5,695
7	33	5,895	610	24	6,783	8,818
10	50	7,735	561	26	8,697	11,306

(*Continued*)

Table 11-30
Heat Pumps (Continued)

Tons of Cooling	Heating Capacity	Direct Material Costs Each	Direct Material Costs Per Ton	Labor Man Hours	Total Material & Labor Direct Costs	Total Material & Labor With 30% O&P
15	64	10,684	516	30	11,794	15,332
25	119	17,312	502	40	18,792	24,430
30	163	24,909	602	44	26,537	34,498
40	193	27,538	499	50	29,388	38,204
Package Unit, Air to Air						
2	6.5	$1,507	$546	5	$1,692	$2,200
3	10	2,167	523	6	2,389	3,105
4	13	2,491	451	8	2,787	3,623
5	27	3,215	466	12	3,659	4,757
7	35	4,524	468	14	5,042	6,554
15	56	11,799	570	18	12,465	16,205
20	100	15,079	546	21	15,856	20,613
25	120	17,961	521	24	18,849	24,503
30	163	22,943	554	26	23,905	31,076
Package Unit, Water Source to Air						
1	13	$1,111	$805	5	$1,296	$1,685
2	19	1,118	405	6	1,340	1,742
3	27	1,511	365	7	1,770	2,301
4	31	1,967	356	8	2,263	2,941
5	29	2,657	385	10	3,027	3,934
7.5	35	3,374	326	16	3,966	5,156
10	50	4,195	304	18	4,861	6,320
15	64	7,335	354	32	8,519	11,074
20	100	8,390	304	36	9,722	12,639

1. Add 5 percent on material for supplementary electric heating coils on air to air units.
2. For water source units add 10 percent to material.

Direct labor costs are $37.00 per hour.

Table 11-31
CONDENSING UNITS
Air Cooled, Staged Compressors, Controls

TONS	CFM	DIRECT MATERIAL COST Each	DIRECT MATERIAL COST Per Ton	LABOR Man Hours	TOTAL MATERIAL & LABOR Direct Cost	TOTAL MATERIAL & LABOR With 30% O&P
3	1,200	$1,676	$559	4	$1,824	$2,372
5	2,000	2,875	575	6	3,097	4,026
7.5	3,000	3,737	498	8	4,033	5,243
10	4,000	4,983	498	9	5,316	6,911
12.5	5,000	5,847	468	10	6,217	8,082
15	6,000	6,708	447	12	7,152	9,298
20	8,000	8,913	446	14	9,431	12,260
25	10,000	11,101	444	14	11,619	15,105
30	12,000	13,225	441	18	13,891	18,059
40	16,000	17,571	439	22	18,385	23,901
50	20,000	21,963	439	24	22,851	29,706
60	24,000	26,260	438	30	27,370	35,581
80	32,000	35,013	438	36	36,345	47,249
100	40,000	43,768	438	42	45,322	58,918

Table 11-32
CHILLED WATER COILS
6 Row with Drain Pan — Water 45°F in, 55°F out, 3 to 4 FPS — Air 45°F in, 55°F out, 400 to 600 FPM

Btuh	Tons	Typical Size	Sq Ft Area @500 Fpm	Gpm	Cfm @400 Cfm Per Ton	Direct Material Cost		Labor	Total Material & Labor	
						Each	Per Sq Ft	Man Hours	Direct Costs	With 30% O&P
60,000	5	36×16	4	12	2,000	$480	$120	4	$628	$817
90,000	7.5	36×24	6	18	3,000	600	100	6	822	1,069
120,000	10	48×24	8	24	4,000	649	81	7	908	1,180
144,000	12	48×30	9.6	28.8	4,800	759	79	8	1,055	1,372
180,000	15	48×36	12	36	6,000	890	74	10	1,260	1,638
240,000	20	48×48	16	48	8,000	983	61	12	1,427	1,855
300,000	25	60×48	20	60	10,000	1,166	58	14	1,684	2,189
450,000	37.5	72×60	30	90	15,000	1,490	50	18	2,156	2,803
600,000	50	96×60	40	120	20,000	2,098	52	22	2,912	3,785
900,000	75	120×72	60	180	30,000	2 691	45	27	3,690	4 797
1,200,000	100	120×96	80	240	40,000	3 202	40	32	4 ,386	5 701
1,500,000	125	120×120	100	300	50,000	3,864	39	37	5,233	6,803
1,800,000	150	144×120	120	360	60,000	4,471	37	40	5,951	7,737
2,400,000	200	150×150	160	480	80,000	5,741	36	48	7,517	9,772

Correction Factors	Material	Labor
1. 2 row coils	.75	.85
2. 4 row coils	.90	.92

Direct labor costs are $37.00 per hour.

Table 11-33
DX EVAPORATOR COILS
6 Row 400 to 600 FPM — Air 75°F in, 55°F out

Btuh	Tons	Typical Size	Sq Ft Area @ 500 Fpm	Cfm @400 Cfm Per Ton	Direct Material Each	Direct Material Per Sq Ft	Labor Man Hours	Total Material Direct Costs	Total Material With 30% O&P
60,000	5	36×16	4	2,000	$529	$132	4	$677	$880
90,000	7.5	36×24	6	3,000	661	110	6	883	1,148
120,000	10	48×24	8	4,000	713	89	7	972	1,264
144,000	12	48×30	9.6	4,80.0	835	87	8	1,131	1,470
180,000	15	48×36	12	6,000	980	82	10	1,350	1,755
240,000	20	48×48	16	8,000	1,081	68	12	1,525	1,982
300,000	25	60×48	20	10,000	1,283	64	14	1,801	2,342
450,000	37.5	72×60	30	15,000	1,639	55	18	2,305	2,997
600,000	50	96×60	40	20,000	2,307	58	22	3,121	4,058
900 000	75	120×72	60	30,000	2,960	49	27	3,959	5,147
1,200 000	100	120×96	80	40 000	3,522	44	32	4,706	6,117
1,500,000	125	120×120	100	50 000	4,250	43	37	5,619	7,305
1,800,000	150	144×120	120	60,000	4,918	41	40	6,398	8,318
2,400,000	200	150×150	160	80,000	6,315	39	48	8,091	10,518

ESTIMATING INSULATION

Insulation used for HVAC floor plan **ductwork** is generally fiberglass blanket with either a RFK or vinyl barrier, with thicknesses of 1, 1 1/2 or 2 inches. Insulation in equipment rooms may be fiberglass blanket, or if required rigid board fiberglass. Outside ductwork is generally covered with rigid board glass mesh and two layers of black mastic. Ductwork insulation costs are based on per square foot.

HVAC **piping** is generally covered with fiberglass preformed with all service jacket (ASJ), fiberglass blanket with ASJ, or foam plastic. Piping insulation costs are based on per linear foot.

Table 11-34
BUDGET ESTIMATING INSULATION
External Wrapping for Ductwork — Per Square Foot

ITEM	THICKNESS	DENSITY	LABOR	MATERIAL COST	SELLING PRICE* w/ O&P
	Inches	Lbs Per Cu Ft	Sq Ft Per Hour	Per Per Sq Ft	Sq Ft
FIBER GLASS BLANKET, RFK					
Vapor Barrier or Vinyl	1	3/4	50	$0.43	$1.80
	1-1/2	3/4	45	0.58	2.20
	2	3/4	40	0.76	2.68
FIBER GLASS, RIGID BOARD					
With RFK Vapor Barrier	1	3	16	0.50	3.98
With 8 oz. Canvas Cover	1	3	15	0.66	4.50
With 8 oz. Canvas and Painted	1	3	14	0.75	4.89
With RFK Vapor Barrier	2	3	13	0.58	4.83
With 8 oz. Canvas Cover	2	3	12	0.54	5.06
With 8 oz. Canvas and Painted	2	3	11	0.60	5.54

Table 11-34 (*Continued*)

OUTSIDE FIBER GLASS, RIGID BOARD					
With 2 Layers black mastic,					
glass mesh	1	4	—	——	4.75
	2	4	—	——	5.29
KITCHEN EXHAUSTS & BREECHING					
Calcium Salicate Block (K Block)					
Rectangular	2	—	8	1.89	9.70
Round	2	—	7	2.54	11.82
With 1/2" Cement					
Rectangular	2	—	5	3.37	16.19
Round	2	—	4	4.43	20.66
Fiber Glass, Rigid Board					
with covering and paint	2	4	8	1.24	8.43

*Includes 15% waste, pins, staples, tape, installation, supervision. 20% overhead and 5% profit. Based on union wages of $37.00 per hour, 10 foot high ducts, lower floors, normal space conditions.

CORRECTION FACTORS

1. Congested ceiling spaces, areas 1.15
2. Wage rate at $37.00 per hour

Table 11-35
PIPING INSULATION
Fiberglass With All Service Jacket (ASJ)
Installed Prices With Overhead And Profit

	THICKNESS			
DIA.	1/2"	1"	1-1/2"	2"
1/2"	3.04	3.29	4.47	5.77
3/4"	3.23	3.54	4.65	6.00
1"	3.38	3.69	4.89	6.38
1-1/4"	3.59	3.94	5.21	6.75
1-1/2"	3.69	4.06	5.35	6.93
2"	3.89	4.31	5.72	7.26
2-1/2"	4.20	4.64	6.14	7.77
3"	4.52	5.03	6.42	8.33
3-1/2"	4.79	5.40	6.93	S.93
4"	5.54	6.19	7.58	9.86
5"	6.05	6.89	8.33	10.89
6"	7.07	7.68	9.27	12.00
7"		8.84	10.33	13.54
8"		9.91	11.40	14.88
9"		10.93	12.47	16.14
10"		11.30	13.49	17.44
12"		12.79	14.56	18.60
14"		13.77	16.23	21.39

Table 11-36
PIPING INSULATION
Calcium Silicate With Cover
Installed Prices With Overhead And Profit

	THICKNESS			
DIA.	1"	1-1/2"	2"	3"
1/2"	5.21	5.82	7.49	
3/4"	5.21	5.86	7.63	
1"	5.12	6.05	7.82	
1-1/4"	5.26	6.38	8.28	10.98
1-1/2"	5.26	6.51	8.47	11.30
2"	5.77	6.89	8.84	11.96
2-1/2"	5.96	7.21	9.77	12.79
3"	6.33	7.49	9.72	12.84
3-1/2"	7.03			
4"	7.49	8.47	10.79	15.35
5"	7.96	9.07	11.86	
6"	5.82	9.63	12.98	17.72
7"				
8"		11.68	17.44	20.46
9"				
10"		14.28	20.46	25.11
12"	17.17	23.25	26.04	
14"	21.39	25.11	29.76	

Table 11-37
PIPING INSULATION
Foam Polyethylene, UV Resistant
Installed Prices With Overhead And Profit

	THICKNESS			
DIA.	3/8"	1/2"	3/4"	1"
1/8"	2.41	2.71	3.02	
1/4"	2.44	2.72	3.04	3.46
3/8"	2.46	2.73	3.11	3.55
1/2"	2.55	2.78	3.26	3.82
3/4"	2.64	2.87	3.37	3.94
1"	2.74	2.99	3.57	4.34
1-1/4"	2.85	3.13	3.85	4.53
1-1/2"	2.90	3.26	3.95	4.58
2"	3.04	3.44	4.4~	5.03
2-1/2"	3.27	3.70	4.74	5.30
3"	3.39	4.01	5.07	5.58
4"		4.49	5.77	

Table 11-38
PIPING INSULATION
Rubber Tubing Foam
Installed Prices With Overhead And Profit

DIA.	THICKNESS 3/8"	1/2"	3/4"
1/4"	2.72	3.67	3.85
3/8"	2.75	3.72	3.93
1/2"	2.89	3.79	4.07
3/4"	2.94	3.85	4.23
1"	3.10	3.93	4.39
1-1/4"	3.19	4.05	4.79
1-1/2"	3.29	4.18	4.93
2"	3.69	4.39	5.26
1/2"	4.17	4.65	5.86
3"	4.30	5.21	6.42
3-1/2"	4.79	5.26	6.70
4"		5.63	7.12
5"		6.51	7.82
6"			10.56

ESTIMATING TEMPERATURE CONTROL WORK

This section covers budget estimating built up control systems in new buildings, compressor systems, control valves and energy management systems.

Table 11-39
BUDGET CONTROL PRICING FOR BUILT-UP SYSTEMS IN NEW BUILDINGS

	Percent of Equipment Costs
Under $25,000	24%
$25,000 to 50.000	20
$60.000 to $100,000	17
$100,000 to $250,000	15
$250,000 to $500,000	12

Add for computerized controls + 10%

Controls for Package Systems	
Under $25,000	8%
$25,000 to $100,000	7
$100,000 to $500,000	6

Table 11-40
PNEUMATIC TEMPERATURE CONTROL COMPRESSOR SYSTEMS INCLUDES COMPRESSOR, TANK, TUBING, DRYER, FILTER, PRV VALVE, MOTOR, STARTER

HP	Direct Material Cost			Labor	Total Material & Labor	
	Duplex Compressor	Other Items	Total Matl Costs	Man Hours	Direct Cost	With 30% O&P
1/2	$1,063	$4,181	$5,244	20	$5,984	$7,779
3/4	1,242	5,520	6,762	21	7,539	9,801
1	1,428	5,679	7,107	23	7,958	10,345
1-1/2	1,750	6,116	7,866	25	8,791	11,428
2	2,022	6,948	8,970	30	10,080	13,104
3	2,342	7,870	10,212	35	11,507	14,959
5	2,625	9,795	12,420	40	13,900	18,070

1. Includes installation of compressor, piping hook up, valves, etc.
2. To furnish and install 3/8" diameter copper trunk line tubing add $6.50 per foot
3. If polyvinyl tubing is used instead of copper tubing deduct 25% from copper tubing.
   Direct labor costs are $37.00 per hour.

Table 11-41
TWO WAY CONTROL VALVES

Dia. Inches	Cv	Direct Costs Each	Type Connect & Press	Labor Man hrs	Total Direct Costs	Matl & Labor With 30% O&P
1/2	.63,1.0,1.6	139.00	Screw, hp	0.55	153.85	200.01
1/2	2.5,4.0	86.00	Screw	0.55	100.85	131.11
3/4	6.3	145.00	Screw,hp	0.60	161.20	209.56
3/4	6.3	89.00	Screw	0.60	105.20	136.76
1	10.0170.00Screw,hp		0.65	187.55	243.82	
1	10.095.00 Screw		0.65	112.55	146.32	
1-1/4	16.0194.00Screw,hp		0.77	214.79	279.23	
1-1/4	16.0117;00Screw		0.77	137.79	179.13	
1-1/2	25.0230.00Screw,hp		0.83	252.41	328.13	
1-1/2	25.0156.00Screw		0.83	178.41	231.93	
2	40.0212.00Screw,hp		0.88	235.76	306.49	
2	40.0201.00Screw		0.88	224.76	292.19	
2-1/2	63.0278.00Screw		2.00	332.00	431.60	
2-1/2	63.0561.00Flange		2.00	615.00	799.50	
3	100.0375.00Screw		2.10	431.70	561.21	

Table 11-41 (*Continued*)
TWO WAY CONTROL VALVES

Dia. Inches	Cv	Direct Costs Each	Type Connect & Press	Labor Man hrs	Total Direct Costs	Matl & Labor With 30% O&P
3	100.0	672.00	Flange	2.10	728.70	947.31
4	160.0	930.00	Flange	2.40	994.80	1293.24

Three Way Control Valves

Mixing

Dia. Inches	Cv	Directs Costs Each	Type Connect	Labor Man hrs	Total Direct Costs	Matl & Labor with 30% O&P
1/2	4.0	112.00	Screw	0.55	126.85	164 91
3/4	6.3	124.00	Screw	0.60	140.20	182 26
1	10.0	144.00	Screw	0.65	161.55	210.02
1-1/4	16.0	169.00	Screw	0.77	189.79	246 73
1-1/2	25.0	200.00	Screw	0.83	222.41	289 13
2	40.0	235.00	Screw	0.88	258.76	336.39
2-1/2	63.0	638.00	Flange	2.00	692.00	899.60

Table 11-41(*Continued*)

Dia. Inches	Cv	Direct Costs Each	Type Connect & Press	Labor Man hrs	Total Direct Costs	Matl & Labor With 30% O&P
3	100.0	755.00	Flange	2.10	811.70	1055.21
4	160.0	1263.00	Flange	2.40	1327.80	1726.14

<u>Diverting</u>

Dia. Inches	Cv	Directs Costs Each	Type Connect	Labor Man hrs	Total Direct Costs	Matl & Labor with 30% O&P
2-1/2	63.0	795.00	Flange	2.00	849.00	1103.70
3	100.0	891.00	Flange	2.10	947.70	1232.01

Not Included:
1. Control motor
2. Linkages
3. Plug in balance relays

hp: High pressure

Direct labor costs are $37.00 per hour.

Table 11-42
ENERGY MANAGEMENT SYSTEMS
(Indirect, Low Voltage, Hardwire)

EQUIPMENT PRICES			
10 Channel EMS unit W7010H*			$4,167
20 Channel EMS unit W7020H*			$6,834
INSTALLED PRICES			
10 Channel @ $1,200/pt			$16,080
20 Channel @ $1,000/pt			$26,800
COMMUNICATIONS LINK HARDWARE			$1,608
REMOTE COMPUTER			
For Programming and Monitoring	$2,680	to	$4,020
(Can put whole program on a floppy disk)			

FUNCTIONS

- Time of Day Scheduling
- Optimum Start/Stop
- Demand Limiting
- Duty Cycling, Temperature Compensated by Temperatures in spaces
- Monitoring Outside Air and Indoor Air Temperatures

*Approximate Honeywell Contractor Price

Direct labor costs are $37.00 per hour.

BUDGET ESTIMATING ENERGY RETROFIT COSTS

Tables on budget costs for various energy retrofit situations, chiller retrofit budget costs and comparisons of retrofit costs to energy savings are covered in this section.

Budget estimating energy retrofit costs involves audits, engineering calculations, financial evaluations and estimating costs to remove and replace HVAC equipment, ductwork, piping, etc. rebalancing, etc.

Budget Estimating Energy Retrofit Costs and Estimating Procedures

COSTS TO REMOVE AND REPLACE

1. New pumps .... $3 to $10/gpm
2. New fan driver & motors .... $100/hp+
3. New chillers .... $200 to $300/ton
4. New boilers .... $6 to $300/ton
5. Heat recovery installations .... $1 to$3/cfm
6. Rebalance .... $10 to $20 outlet
   .... $60 to $120 per ahu coil
   .... $20 per terminal unit
7. Convert CAV to VAV .... $400 to $1200/box
   .... $.35 to $1/sq. ft
8. Audits .... 2¢ to 5¢/sq. ft.
9. Overall energy retrofit programs .... $.25 to $4/sq. ft.

Table 11-43
BUDGET ESTIMATING ENERGY RETROFIT COSTS
Compared to Btu and Costs Savings

Retrofit Investment Cost Per Sq Ft Building	Rough Yearly Btu's Saved Per Sq Ft	Yearly Energy Costs Saved Dollars		ROI Percent In Decimals
		Per Sq Ft	Per 1,000 Sq Ft	
Office Buildings				
$0.25	40,000	$0.86	$860.00	3.37
0.50	46,000	0.99	990.00	1.91
0.75	50,000	1.08	1,080.00	1.37

(Continued)

Table 11-43 (*Continued*)

Retrofit Investment Cost Per Sq Ft Building	Rough Yearly Btu's Saved Per Sq Ft	Yearly Energy Costs Saved Dollars		ROI Percent In Decimals
		Per Sq Ft	Per 1,000 Sq Ft	
<u>Office Buildings</u>				
1.00	53,000	1.14	1,140.00	1.07
1.25	56,000	1.20	1,200.00	0.89
1.50	58,000	1.25	1,250.00	0.77
1.75	59,000	1.27	1,270.00	0.66
2.00	60,000	1.29	1,290.00	0.58
2.25	60,500	1.30	1,300.00	0.51
2.50	61,000	1.31	1,310.00	0.46
2.75	61,500	1.32	1,320.00	0.41
3.00	62,000	1.33	1,330.00	0.38
Hospitals				
$0.25	45,000	$0.97	$970.00	3.81
0.50	56,000	1.20	1,200.00	2.33
0.75	62,000	1.33	1,330.00	1.71
1.00	65,000	1.40	1,400.00	1.33
1.25	67,000	1.44	1,440.00	1.09
1.50	70,000	1.51	1,510.00	0.94
1.75	72,000	1.55	1,550.00	0.82
2.00	74,000	1.59	1,590.00	0.73
2.25	75,000	1.61	1,610.00	0.65
2.50	76,000	1.63	1,630.00	0.59
2.75	77,000	1.66	1,660.00	0.54
3.00	78,000	1.68	1,680.00	0.49
Universities				
$.025	30,000	$0.65	$650.00	2.53
0.50	52,000	1.12	1,120.00	2.17

(*Continued*)

Table 11-43 (*Continued*)

0.75	61,000	1.31	1,310.00	1.68
1.00	67,000	1.44	1,440.00	1.37
1.25	72,000	1.55	1,550.00	1.17
1.50	77,000	1.66	1,660.00	1.04
1.75	80,000	1.72	1,720.00	0.92
2.00	84,000	1.81	1,810.00	0.84
2.25	87,000	1.87	1,870.00	0.76
2.50	90,000	1.94	1,940.00	0.71
2.75	92,000	1.98	1,980.00	0.65
3.00	93,000	2.00	2,000.00	0.60

1. Total savings and ROI based on electrical and fuel consumption.
2. Electrical savings based on 66 percent of total consumption.
3. Gas and oil based on 34 percent of total consumption.

Table 11-44
BUDGET ESTIMATING CHILLER RETROFITS
300-Ton Chiller

	Material Costs	Labor Hours	Totals w/30% MU
Chiller @ $200/ton	$60,000	130	$75,000
Piping	$1,000	100	$5,000
Electrical	$1,000	100	$5,000
Insulation	$1,000	100	$5000
Controls	$1,000	100	$5,000
General Construction	$1,000	100	$5,000
Total	$65,000	100	$5,000

Hence $100,000 divided by 300 tons = $333 per ton

Table 11-45
ESTIMATING AIR TESTING AND BALANCING WORK

	New System		Rebalance	
	Hr Range	Avg Cost w/ O & P	Hr Range	Avg Cost w/ O & P
Check Out and Test Equipment				
Built up Supply Unit	4.0 - 6.0	$200	3.0 - 5.0	$160
Air Handling Unit	3.0 - 5.0	160	2.0 - 4.0	120
Roof Top Unit, SZ	2.0 - 3.0	100	1.5 - 2.5	80
Roof Top Unit, MZ	2.0 - 4.0	120	1.5 - 3.0	100
plus per zone	.5	20	.4	16
Fan-Coil units under windows	0.5 - 1.0	30	0.5 - 1.0	30
Centrifugal Fans	2.0 - 3.0	100	1.0 - 2.0	60
Roof Exhaust Fans	1.0 - 2.0	60	0.5 - 1.0	30
Dust Collectors	2.0 - 4.0	120	1.0 - 3.0	80
Balance Outlets (No equipment time included)				
Supply Diffusers: Flow Hood	.25 -.35	12	.20 -.25	10
Other	.35 -.50	17	.25 -.35	12
Linear Diff.,Per 4ft: Flow Hood	.25 -.35	12	—	10
Air/Light Troffers: Flow Hood	.25 -.35	12	—	10
Small Registers: Flow Hood	.20 -.30	10	—	8
Other	.30 -.40	14	—	12
Large Registers: Flow Hood	.25 -.35	12	—	12
Other	.35 -.50	17	—	17
Small Exhaust Hoods:	0.4 - 0.6	20	—	—
Large Exhaust Hoods:	0.6 - 0.8	28	—	—
Air Terminal Units, CAV and VAV				
Above Ceiling: (Check every 3rd unit)	0.4 - 0.6	20	—	—
Balance Window Induction Units:	0.5 - 0.724	—	—	

(*Continued*)

Table 11-45 (*Continued*)

New System	Hr Range	Avg Cost w/ O & P	Rebalance Hr Range	Avg Cost w/ O & P

Labor Correction Factors (Multipliers)
1. Small, simple systems, open areas.................................0.75
2. Large, complicated systems, many separate areas......1.25

NOTES
1. Labor costs with overhead and profit, $40.00 per hour.
2. Rebalance time, 60 to 80 percent of new balance time.
3. Abbreviations; SZ: Single Zone MZ: Multi-Zone Other: Other type instrument, anomometer, hot wire, etc.

Figure 11-5
ESTIMATING DUCTWORK LEAK TESTING
Medium Or High Pressure Duct Runs

LABOR FOR TYPICAL DUCT RUN

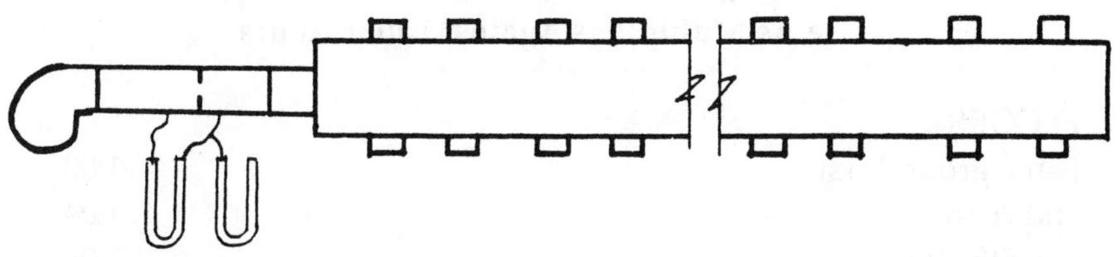

1.	Set up leak testing rig	1.0 hr
2.	Cap and seal ends of duct run, 1/2 hr each	1.0 hr
3.	Cap and seal branch collars, 16 @.2 hr each	3.2 hr
4.	Check leakage with fan, walk run, seal	2.0 hr
5.	Retest	1.0 hr
	Total time for segment	8.2 hrs.

Table 11-46
ESTIMATING HYDRONIC TESTING AND BALANCING

Check Out Equipment	Hours
Pumps: Check out pump itself, motor, starter, adjacent valves; read pressures, gpm, amps, volts; adjust.	1.5 to 2.5
Chillers, Absorption Units	2.0 to 3.0
Cooling Tower	5.0 to 10.0
Central Cooling and Heating Coils	2.0 to 4.0
Balancing Terminals	
Reheat Coils, Radiation Units	.75 to 1.25
Induction Units, Fan Coil Units	.75 to 1.25

Table 11-47
FIELD LABOR CORRECTION FACTORS
FOR HVAC EQUIPMENT, DUCTWORK & PIPING
Use As Multipliers Against Labor Hours

FLOORS	
Bsmt, ground, 1st	1.00
2nd & 3rd	1.04
4th, 5th, 6th	1.08
7th, 8th, 9th	1.12
10th, 11th, 12th	1.16
13th, 14th, 15th	1.21
16th, 17th, 18th	1.26
19th, 20th, 21st	1.31

(Continued)

Table 11-47 (*Continued*)

DUCT OR PIPING HEIGHTS ABOVE FLOOR

10 ft	1.00
15 ft	1.10
20 ft	1.20
25 ft	1.30
30 ft	1.40
35 ft	1.50
40 ft	1.60

CORRECTION FOR SIZE OF JOB

0 - 24 hrs	1.12
25 - 48 hrs	1.08
48 - 96 hrs	1.04
96 hrs up	none

SPECIAL AREAS

Open areas, no partitions	.85
Congested ceiling space	1.15
Equipment room	1.20
Kitchen	1.10
Auditorium; pool w/sloped floor	1.25
Attic space	1.50
Crawl space	1.20
Cramped shaft	1.30
One or two continuous risers	.80

DISTANCE FROM UNLOADING POINT

100 ft	1.00
200 ft	1.03
300 ft	1.05

EXISTING BUILDING

Typical existing office bldg, hospital, school, etc.	1.35
Existing factory, warehouse, gym, hall, garage, no ceilings	1.15
100% gutted area	1.10

(*Continued*)

Table 11-47 (*Continued*)

Work around, over, machinery, furniture	1.10
Protect machinery, floor furniture	1.05
Quiet job	1.05
Occupied areas	1.05
Remove items and reinstall same	1.50
Remove and replace equipment	2.00

EXISTING BUILDING CONSIDERATIONS

Working on HVAC systems in existing buildings takes more time than it does in new buildings. It involves more complications and requires that many additional factors be considered when estimating or scheduling than as with new building work.

OPENINGS TO CUT	Do openings need to be cut? Do they have to be patched? What's the material and thickness of walls, roofs or floors to be cut? Quantity and sizes of openings?
CEILINGS TO REMOVE	Do ceilings have to be removed? Who replaces ceilings? Entire ceilings being removed or just cut where needed? What type of ceilings are involved and how high are they?
REMOVAL OF EXISTING HVAC	Fans, pumps, HVAC units, ductwork, EQUIPMENT, ETC. piping, etc?
PROTECTION	Does owners furniture or equipment have to be protected?

CEILING SPACE AVAILABLE	What ceiling space is available? Where are beams located?
EQUIPMENT ROOM SPACE AVAILABLE	Is there space for the new HVAC equipment, ductwork, piping, etc?
OBSTRUCTIONS TO AVOID	Piping and ducts Lights and conduit Walls and partitions Beams, joists and columns
OCCUPIED AREAS	Will areas be occupied during installation?
SHUTTING DOWN OF SYSTEMS	Must systems be in operation or can they be shut down during installation?
SEQUENCE OF WORK	In what order must work be done?
WORKING TIMES	Week days, evenings, weekends. Over time
MATERIAL HANDLING AND HOISTING	Elevators available Dock available Door, window, wall openings available and large enough for moving equipment through
CLEAN UP AND SCRAPING	Material handling labor, dumpster rentals, scavenger services.

Chapter 12

RECAP

Variable air volume systems vary air flow to the spaces at a constant temperature to meet the variable heating and cooling loads. Constant air volume systems supply a constant volume of air, but vary the temperature to meet the heat and cooling loads of the spaces. VAV systems may cost more initially to install, but they can save 20 to 30 percent in operational costs over constant air volume systems.

The general objectives of an HVAC system are to provide proper comfort and health conditions for occupants, meet code requirements, provide the correct environment for operations and to do so at reasonable first costs and operational costs.

The HVAC system does this by controlling temperatures, humidity, air speeds, air currents, air cleanliness, noise levels and building pressures.

Hence, the VAV system, in addition to maintaining correct space cooling temperatures between maximum and minimum air flows, must simultaneously provide heat when needed for correct space heating, control air speeds and currents in the spaces, meet humidity requirements, maintain building and space pressures, etc., throughout the maximum and minimum air flow ranges.

To meet all these HVAC goals there is a large selection of types of VAV systems, terminals, fan volume controls, sensors and controls available.

TYPE SYSTEMS

For smaller systems in the 5- to 50-ton range, lower priced secondary systems can be used. These systems may or may not save energy costs, but they do provide space load control by varying the air flow to different zones. Secondary VAV systems usually have pressure dependent terminals, no special fan volume controls and no main duct static pressure or volume sensors. Typical secondary systems are terminal or fan bypass types, variable volume/variable temperature damper types, riding the fan curve and system powered systems.

For larger systems over 50 or 60 tons, true variable air volume systems are used with pressure independent terminals, fan volume controls and trunk duct pressure or volume sensing. Typical true VAV systems are cooling-only, cooling with reheat, separate interior and exterior system, fan powered, dual duct and induction.

TERMINALS

There are also many different types of VAV terminals available. Terminals are classified by their type of control power, whether they are pressure dependent or independent, and whether they are normally open or closed. The components of a VAV box consist of the housing, inlet, discharge, damper, differential sensor and volume controller device.

Smaller VAV systems use pressure dependent cooling-only, reheat, damper, bypass, system powered and volume limiting type terminals.

Larger VAV systems use pressure independent cooling-only, reheat, fan powered, dual duct and induction type terminals.

Terminals must be properly calibrated and set for specific maximum and minimum flows.

FAN VOLUME CONTROL METHODS

There are a number of different methods available for varying fan volume to meet the varying terminal CFM requirements such as inlet or discharge dampers, motor inverters, automatic variable pitch sheaves, eddy current couplings, etc.

VAV CONTROLS

VAV control systems can be powered pneumatically, electrically or with DDC. The basic components of a VAV control system consist of temperature, pressure and humidity sensors, dampers and valves, air flow monitors, control motors, transducers, square root extractors, etc.

Not only must the flow, temperature and humidity be controlled, but provisions have to be made to control building pressures, economy cycles, night heating and morning warmups and proper tracking of the supply air by the return fan.

TESTING AND BALANCING VAV SYSTEM

When testing and balancing a VAV system it is set to maximum flow and tested as a constant volume system. The system must be checked out at minimum flow also. If there is diversity, which is where the maximum flow from the fan is less than the total of all the terminals, the whole system must be tested in segments.

MORE SENSITIVE AREAS OF VAV SYSTEMS

Some of the more difficult areas in VAV systems are maintaining proper building and space pressures, controlling the outside air volume, controlling return fan volume and return air fan tracking of the supply fan, throughout the entire swing from maximum to minimum flow.

Diversity can be a problem if actual building loads exceed design loads.

The air flow out of supply outlets can be a problem due to being improperly selected or sized.

Chapter 13
APPENDICES
Appendix A — Abbreviations & Symbols

A Area, Square Feet
A Amps
act Actual

BHP Brake Horsepower
Btu British Thermal Unit
BtuH British Thermal Unit per Hour

CFM Cubic Feet per Minute
Cu Cubic
Cu ft Cubic Feet
CHW Chilled Water
CHWS Chilled Water Supply
CHWR Chiller Water Return

D Difference, Delta
DB Dry Bulb Temperature
DIA, dia Diameter
DP, ΔP Difference in Pressure
Dt, Δt Difference in Temperature

E Voltage
eff Efficiency

F Fahrenheit
fpm Feet per Minute
ft Feet

GPH Gallons per Hour
GPM Gallons per Minute

H Head
Hf Friction Head
HP High Pressure System
Hs Elevation Head
HW Hot Water
HWS Hot Water Supply
HWR Hot Water Return

I Amperes

KVA Kilovolt-Amperes
kW Kilowatts

LP Low Pressure System

MP Medium Pressure System
MZ Multi-Zone

NPSH Net Positive Suction Heat

Ns Specific Speed
OA Outside Air

P Pressure
Pa Atmospheric or Absolute Pressure
PF Power Factor

Pvp	Vapor Pressure
R	Rankine Temperature
R	Resistance, Ohms
RA	Return Air
RH	Relative Humidity
rpm	Revolutions per Minute
SP	Static Pressure, Inches Water Gauge
Sp gr	Specific Gravity
sq ft	Square Feet
SZ	Single Zone System
t	Temperature, degrees Fahrenheit or Celcius
T	Absolute Temperature, 460 + Degrees Fahrenheit
TP	Total Pressure, Inches Water
V	Velocity, Feet per Minute
V	Volts
VP	Velocity Pressure, Inches Water Gauge
WB	Wet Bulb Temperature
WG	Water Gauge

Appendix B
ASSOCIATION ABBREVIATIONS

ACCA	Air Condition Contractors Association of America
ADC	Air Diffusion Council
ADI	Air Distribution Institute
AMCA	Air Moving and Control Association, Inc.
ARI	Air Conditioning and Refrigeration Institute
ASHRAE	American Society of Heating, Refrigeration and Air Conditioning Engineers, Inc.
ASME	American Society of Mechanical Engineers
MCAA	Mechanical Contractors Association of America, Inc.
NAPHCC	National Association of Plumbing-Heating-Cooling Contractors
NEMA	National Electrical Manufacturers Association
NSPE	National Society of Professional Engineers
SMACNA	Sheetmetal and Air Conditioning Contractors National Association, Inc.

Appendix C
AIR FLOW AND AIR PRESSURE FORMULAS

- **AIR FLOW FORMULA**

Used to find volume of air flowing through ductwork, outlets, inlets, hoods, etc.

Basic Formula $CFM = A \times V$ where:

CFM	=	cubic fpm
A	=	area in sq ft
V	=	velocity in fpm
A_k	=	factor used with outlets; actual unobstructed air flow area

Velocity unknown:

$$V = \frac{CFM}{A}$$

Area unknown:

$$A = \frac{CFM}{V}$$

- **TOTAL PRESSURE FORMULA**

Measure of total pressure energy in air at any particular point in an air distribution system.

$$TP = VP + SP$$

rearranged:

$$VP = TP - SP$$

$$SP = TP - VP$$

where:

TP	=	total pressure inches W.G.
VP	=	velocity pressure, inches W.B.
SP	=	static pressure, inches W.G.

- **CONVERTING VELOCITY PRESSURE INTO FPM**

Standard Air, 075 lb/cu ft

$$fpm = 4005 \times \sqrt{VP}$$

rearranged:

$$VP = \left(\frac{fpm}{4005}\right)^2$$

Non Standard Air

$$fpm = 1096 \times \sqrt{\frac{VP}{density}}$$

where:

VP	=	velocity pressure, inches W.G.
density	=	lb/cu ft
fpm	=	feet per minute

Appendix D
CHANGING FAN CFMs AND DRIVES

- **FAN LAW NO. 1**

$$\text{RPM new} = \text{RPM old} \times \left(\frac{\text{CFM new}}{\text{CFM old}}\right)$$

$$\text{SP new} = \text{SP old} \times \left(\frac{\text{CFM new}}{\text{CFM old}}\right)^2$$

$$\text{BHP new} = \text{BHP old} \times \left(\frac{\text{CFM new}}{\text{CFM old}}\right)^3$$

where:

RPM	=	revolutions per minute
CFM	=	cubic feet per minute
SP	=	fan static pressure, inches W.G.
BHP	=	brake horsepower

- **BHP FORMULAS**

$$\underset{\text{(3 phase)}}{\text{BHP actual}} = \frac{1.73 \times \text{amps} \times \text{volts} \times \text{eff.} \times \text{power factor}}{746}$$

$$\underset{\text{(rule of thumb)}}{\text{BHP actual}} = \begin{matrix}\text{(name plate)}\\ \text{horsepower}\end{matrix} \times \left(\frac{\text{amps act.}}{\text{amps rated}}\right) \times \left(\frac{\text{volts act.}}{\text{volts rated}}\right)$$

where:

BHP = brake horsepower

eff = efficiency

- **SHEAVE/RPM RATIOS & BELT LENGTHS**

$$\frac{\text{RPM motor}}{\text{RPM fan}}\left(\begin{matrix}\text{Speed}\\\text{Ratio}\end{matrix}\right) = \frac{\text{DIA fan sheave}}{\text{DIA motor sheave}}\left(\begin{matrix}\text{Diameter}\\\text{Ratio}\end{matrix}\right)$$

$$\text{DIA fan sheave} = \text{DIA motor sheave} \times \left(\frac{\text{RPM motor}}{\text{RPM fan}}\right)$$

$$\text{DIA motor sheave} = \text{DIA fan sheave} \times \left(\frac{\text{RPM fanr}}{\text{RPM motor}}\right)$$

$$\text{Belt Length} = 2c + \left[1.57 \times (D+d)\right] + \frac{(D-d)2}{4c}$$

where:

C = center to center distance of shaft

D = large sheave diameter

d = small sheave diameter

Appendix E
AIR HEAT TRANSFER FORMULAS

- **SENSIBLE**

  BtuH = CFM x temp change x 1.08

  Rearranged:

  $$\text{CFM} = \frac{\text{BtuH (Sensible}}{1.08 \times \text{temp change}}$$

  Rearranged:

  $$\text{Temp Change} = \frac{\text{BtuH (Sensible}}{\text{CFM} \times 1.08}$$

- **LATENT**

  BtuH = 4840 x CFM x WD

- **TOTAL LATENT AND SENSIBLE**

  BtuH = 4.5 x CFM x HD

where:

BtuH = British thermal units per hour
T = Temperature, F
CFM = Cubic feet per minute
EFF = Efficiency

HD = Difference in enthalpy
WD = Difference in humidity ratio (lb water/lb dry air)

- **AIR FLOW FOR FURNACES**

Gas Furnace

$$CFM = \frac{\text{Heat value of gas (Btu/cu ft) x cu ft/hr x 0.75}}{\text{1.08 x Temp Rise*}}$$

Oil Furnace

$$CFM = \frac{\text{Heat value of oil (Btu/gal) x gal/hr x 0.70}}{\text{1.08 x Temp Rise*}}$$

Electric Furnace

$$CFM = \frac{\text{Volts x Amps x 3.4}}{\text{1.08 x Temp Rise*}}$$

where:

BtuH = British thermal units per hour
RH = relative humidity, percent
T = temperature, °F
CFM = cubic feet per minute
* Difference between supply air and return air temperatures

Appendix F
CHANGES IN STATE OF AIR FORMULAS

For changes of state in volume, temperature and pressure of air and other gases.

New volume of air pressure remains constant and temperature changes

$$V_{new} = V_{orig} \times \left(\frac{T_{new}}{T_{orig}}\right)$$

New volume when temperature remains constant and pressure changes

$$V_{new} = V_{orig} \times \left(\frac{P_{orig}}{P_{new}}\right)$$

New pressure when volume is constant and temperature changes

$$P_{new} = P_{old} \times \left(\frac{T_{new}}{T_{old}}\right)$$

Derived from gas laws:

$$PV = mrT$$

$$\left(\frac{PV}{T}\right)_{\text{orig state}} = \left(\frac{PV}{T}\right)_{\text{new state}}$$

where:

V	=	volume in cu ft
P	=	absolute pressure (atmospheric + gauge pressure)
T	=	absolute temperature (460 + °F)

Appendix G

AIR DENSITY CORRECTION FACTORS

For Different Altitudes and Temperatures

Air Temp. °F	Sea Level	1000	2000	3000	4000	5000	6000	7000	8000	9000	10,000
	Altitude (ft)										
−40°	1.26	1.22	1.17	1.13	1.09	1.05	1.01	0.97	0.93	0.90	0.87
0°	1.15	1.11	1.07	1.03	0.99	0.95	0.91	0.89	0.85	0.82	0.79
40°	1.06	1.02	0.99	0.95	0.92	0.88	0.85	0.82	0.79	0.76	0.73
70°	1.00	0.96	0.93	0.89	0.86	0.83	0.80	0.77	0.74	0.71	0.69
100°	0.95	0.92	0.88	0.85	0.81	0.78	0.75	0.73	0.70	0.68	0.65
150°	0.87	0.84	0.81	0.78	0.75	0.72	0.69	0.67	0.65	0.62	0.60
200°	0.80	0.77	0.74	0.71	0.69	0.66	0.64	0.62	0.60	0.57	0.55
250°	0.75	0.72	0.70	0.67	0.64	0.62	0.60	0.58	0.56	0.58	0.51
300°	0.70	0.67	0.65	0.62	0.60	0.58	0.56	0.54	0.52	0.50	0.48
350°	0.65	0.62	0.60	0.58	0.56	0.54	0.52	0.51	0.49	0.47	0.45
400°	0.62	0.60	0.57	0.55	0.53	0.51	0.49	0.48	0.46	0.44	0.42
450°	0.58	0.56	0.54	0.52	0.50	0.48	0.46	0.45	0.43	0.42	0.40
500°	0.55	0.53	0.51	0.49	0.47	0.45	0.44	0.43	0.41	0.39	0.38
550°	0.53	0.51	0.49	0.47	0.45	0.44	0.42	0.41	0.39	0.38	0.36
600°	0.50	0.48	0.46	0.45	0.43	0.41	0.40	0.39	0.37	0.35	0.34
700°	0.46	0.44	0.43	0.41	0.39	0.38	0.37	0.35	0.34	0.33	0.32
800°	0.42	0.40	0.39	0.37	0.36	0.35	0.33	0.32	0.31	0.30	0.29
900°	0.39	0.37	0.36	0.35	0.33	0.32	0.31	0.30	0.29	0.28	0.27
1000°	0.36	0.35	0.33	0.32	0.31	0.30	0.29	0.28	0.27	0.26	0.25

Standard Air Density, Sea Level, 70°F = 0.075 lb/cu ft

Example:

Find actual density of air at sea level if the temperature is 600°F.

Actual Density = (0.75 lb/cu ft) x (0.52) = .04 lb/cu ft

Find the actual density of 70°F air at 5000 feet altitude.

Actual Density = (0.75 lb/cu ft) x (0.83) = .0622 lb/cu ft

Appendix H
CONVERTING VELOCITY PRESSURE INTO FEET PER MINUTE (Standard air)

Velocity Pressure	FPM
.001	127
.005	283
.010	400
.015	491
.020	566
.025	633
.030	694
.035	749
.040	801
.045	849
.050	896
.055	939
.060	981
.065	1020
.070	1060
.075	1097
.080	1133
.085	1167
.090	1201
.095	1234
.100	1266
.105	1298
.110	1328
.115	1358
.120	1387
.125	1416
.130	1444
.135	1471
.140	1498
.145	1525
.150	1551
.155	1577
.160	1602
.165	1627
.170	1651
.175	1675
.180	1699
.185	1723
.190	1746
.195	1768
.200	1791

Velocity Pressure	FPM
.21	1835
.22	1879
.23	1921
.24	1962
.25	2003
.26	2042
.27	2081
.21	2119
.29	2157
.30	2193
.31	2230
.32	2260
.33	2301
.34	2335
.35	2369
.36	2403
.37	2436
.38	2469
.39	2501
.40	2533
.41	2563
.42	2595
.43	2626
.44	2656
.45	2687
.46	2716
.47	2746
.48	2775
.49	2804
.50	2832
.51	2860
.52	2888
.53	2916
.54	2943
.55	2970
.56	2997
.57	3024
.58	3050
.59	3076
.60	3102

Velocity Pressure	FPM
.61	3127
.62	3153
.63	3179
.64	3204
.65	3229
.66	3254
.67	3279
.68	3303
.69	3327
.70	3351
.71	3375
.72	3398
.73	3422
.74	3445
.75	3468
.76	3491
.77	3514
.78	3537
.79	3560
.80	3582
.81	3604
.82	3625
.83	3657
.84	3669
.85	3690
.86	3709
.87	3729
.88	3758
.89	3779
.90	3800
.91	3821
.92	3842
.93	3863
.94	3884
.95	3904
.96	3924
.97	3945
.98	3965
.99	3985
1.00	4005

Velocity pressure in inches WG
Based on formula: $FPM = 4005\sqrt{VP}$

Appendix I
CONVERTING VELOCITY PRESSURE INTO FEET PER MINUTE
For Various Temperatures

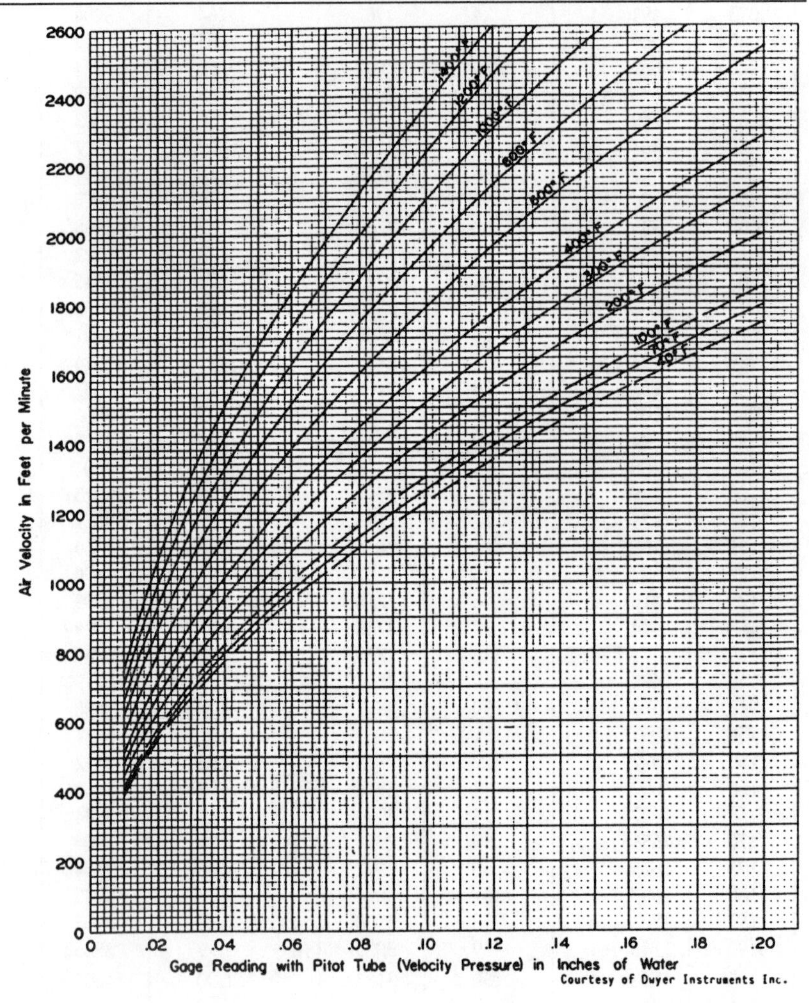

Courtesy of Dwyer Instruments Inc.

Appendix J
HYDRONIC FORMULAS

- **CONVERTING PSI TO FEET OF HEAD:**

ft hd = 2.31 x psi inches hd = 27.2 x psi
psi = .433 x ft hd, psi = .036 x inches hd

- **WATER HEAT TRANSFER FORMULAS**

BtuH = GPM x (T in – T out) x 500

$$GPM = \frac{BtuH}{(T\ in - T\ out) \times 500}$$

- **ELECTRICAL POWER CONSUMPTION OF WATER PUMP**

$$BHP = \frac{GPM \times ft\ head}{3960 \times eff}$$

- **USING SYSTEM COMPONENT AS FLOW MEASURING DEVICE**

$$GPM\ actual = GPM\ design \times \sqrt{\frac{\Delta P\ actual}{\Delta P\ design}}$$

$$\Delta P\ actual = \Delta P\ design \times \left(\frac{GPM\ actual}{GPM\ design}\right)^2$$

- **COIL OR CHILLER GPM**

$$GPM = \frac{Tons \ x \ 24}{T\ in \ - \ T\ out}$$

- **CONDENSER GPM**

$$GPM = \frac{Tons \ x \ (kW \ x \ .284)}{T\ out - T\ in}$$

where:

GPM	=	gallons per minute
Δ P	=	change in pressure across component
BtuH	=	British thermal units per hour
T	=	Temperature, °F
BHP	=	brake horsepower
kW	=	Kilowatts

Appendix K
PSYCHROMETRIC CHART (Sea Level)

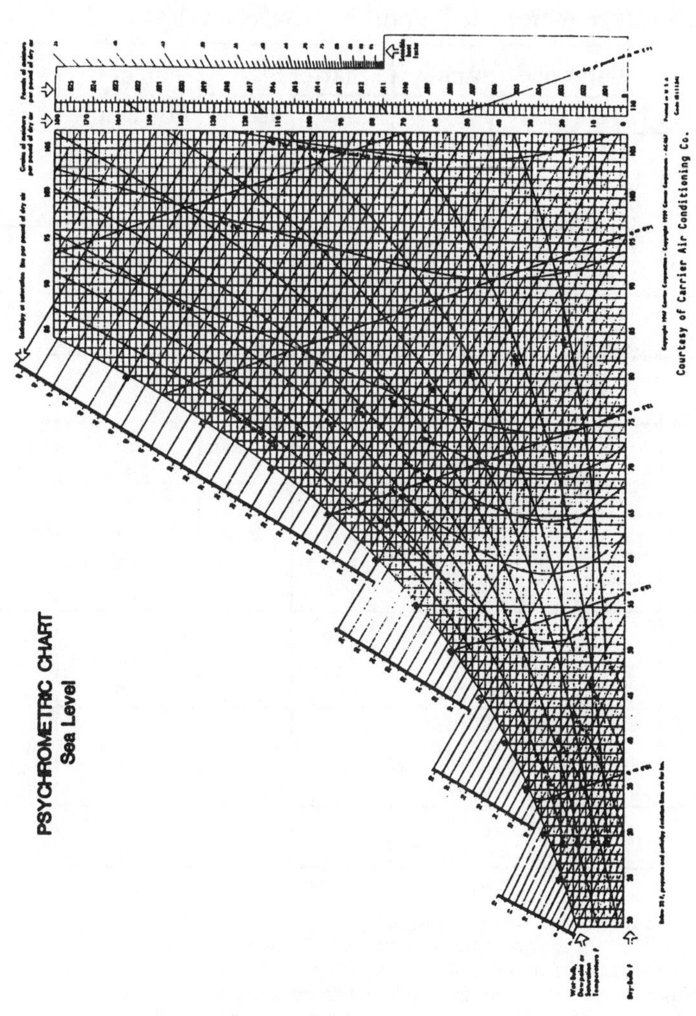

Appendix L
MOTOR AMP DRAWS, EFFICIENCIES, POWER FACTORS, STARTER SIZES
Approximate Values

Induction type motors, 1800 rpm, 3-phase, 60-cycle.

HP	Full Load Amps			Starter Size		Percent	Power
	115V	230V	460V	230V	460V	Efficiency	Factors
1/2	4.0	2.0	1.0	00	00	70	69.2
3/4	5.6	2.8	1.4	00	00	72	72.0
1	7.2	3.6	1.8	0	0	79	76.5
1-1/2	10.4	6.8	3.4	0	0	80	80.5
2	13.6	6.8	3.4	0	0	80	85.3
3	19.2	9.6	4.8	0	0	81	82.6
5	30.4	15.2	7.6	0	0	83	84.2
7-1/2		22	11	1	0	85	85.5
10		28	14	1	1	85	88.8
15		42	21	2	1	86	87.0
20		54	27	2	1	87	87.2
25		68	34	2	2	88	86.8
30		80	40	3	2	89	87.2
40		104	52	3	2	89	88.2
50		130	65	4	3	89	89.2
60		154	77	4	3	89	89.5
75		192	96	4	4	90	89.5
100		248	124	5	4	90	90.3
125		293	147			90	90.5
150		348	174			91	90.5
200		452	226			91	
250		568	284			91	
300		678	339			92	

Rule of Thumb on AMP Draws for Motors

- a 115V motor draws double the amps of a 230V motor.
- A 230V motor draws double the amps of a 460V motor.
- Single-phase motors draw double the amps of 3-phase motors.
- At 115 volts, a 3-phase motor draws about 5.2 amps per HP.
  At 230 volts, a 3-phase motor draws about 2.6 amps per HP.

Electrical Formulas

To Find	Alternating Current	
	Single-phase	Three-phase
Amperes =	$\frac{Hp \times 746}{E \times Eff \times pf}$	$\frac{Hp \times 746}{1.73 \times E \times Eff \times pf}$
Amperes =	$\frac{kW \times 1000}{E \times pf}$	$\frac{kW \times 1000}{1.73 \times E \times ph}$
Amperes =	$\frac{Kva \times 1000}{E}$	$\frac{Kva \times 1000}{1.73 \times E}$

where:

I	=	Current in amps	kWh	=	Kilowatt hours
E	=	Voltage	BHP	=	Brake horsepower
PF	=	Power factor	KVA	=	Kilovolt amps
effM	=	efficiency of motor	HR	=	Hours per year

- **THREE-PHASE, ALTERNATING CURRENT MOTORS**

$$\text{kW actual (Motor Input)} = \frac{1.73 \times I \times E \times PF}{1000}$$

$$\text{BHP (Motor Outout)} = \frac{1.73 \times I \times E \times effM \times PF}{746}$$

$$\text{kWh (Motor Input)} = \frac{BHP}{effM}$$

- **MOTOR ELECTRICAL COSTS FOR YEAR**

$$\text{Paid kWh Input per Year} = \frac{1.73 \times I \times E \times HR}{1000}$$

$$\text{kW Costs per Year} = \frac{1.73 \times I \times E \times HR \times \$kWh}{1000}$$

NEMA MOTOR FRAME DIMENSION STANDARDS

Standardized motor dimensions as established by the National Electrical Manufacturers Association (NEMA) are tabulated below and apply to all base-mounted motors listed herein which carry a NEMA frame designation.

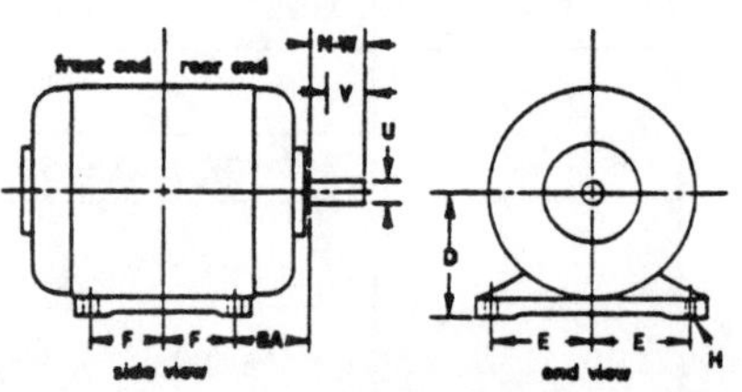

NEMA FRAME	D(*)	2E	2F	BA	H	N-W	U	V(§) Min.	Key Wide	Key Thick	Key Long	NEMA FRAME
	—All Dimensions in Inches—											
42	2 5/8	3 1/2	1 11/16	2 1/16	9/32 slot	1 1/8	3/8	—	—	3/64 flat	—	42
48	3	4 1/4	2 3/4	2 1/2	11/32 slot	1 1/2	1/2	—	—	3/64 flat	—	48
56 56H	3 1/2	4 7/8	3 3&5(‡)	2 3/4	11/32 slot	1 7/8(†)	5/8(†)	—	3/16(†)	3/16(†)	1 3/8(†)	56 56H
56HZ	3 1/2	**	**	**	**	2 1/4	7/8	2	3/16	3/16	1 3/8	56HZ
66	4 1/8	5 7/8	5	3 1/8	13/32 slot	2 1/4	3/4	—	3/16	3/16	1 7/8	66
143T 145T	3 1/2	5 1/2	4 5	2 1/4	11/32 dia.	2 1/4	7/8	2	3/16	3/16	1 3/8	143T 145T
182 184	4 1/2	7 1/2	4 1/2 5 1/2	2 3/4	13/32 dia.	2 1/4	7/8	2	3/16	3/16	1 3/8	182 184
182T 184T			4 1/2 5 1/2			2 3/4	1 1/8	2 1/2	1/4	1/4	1 3/4	182T 184T
213 215	5 1/4	8 1/2	5 1/2 7	3 1/2	13/32 dia.	3	1 1/8	2 3/4	1/4	1/4	2	213 215
213T 215T			5 1/2 7			3 3/8	1 3/8	3 1/8	5/16	5/16	2 3/8	213T 215T
254U 256U	6 1/4	10	8 1/4 10	4 1/4	17/32 dia.	3 3/4	1 3/8	3 1/2	5/16	5/16	2 3/4	254U 256U
254T 256T			8 1/4 10			4	1 5/8	3 3/4	3/8	3/8	2 7/8	254T 256T
284U 286U	7	11	9 1/2 11	4 3/4	17/32 dia.	4 7/8	1 5/8	4 5/8	3/8	3/8	3 3/4	284U 286U
284T 286T			9 1/2 11			4 5/8	1 7/8	4 3/4	1/2	1/2	3 1/4	284T 286T
324U 326U	8	12 1/2	10 1/2 12	5 1/4	21/32 dia.	5 3/8	1 7/8	5 3/8	1/2	1/2	4 1/4	324U 326U
324T 326T			10 1/2 12			5 1/4	2 1/8	5	1/2	1/2	3 7/8	324T 326T
326TS			12			3 3/4 (▲)	1 7/8 (▲)	3 1/2 (▲)	1/2	1/2	2 (▲)	326TS
364U 365U	9	14	11 1/4 12 1/4	5 7/8	21/32 dia.	6 3/8	2 1/8	6 1/8	1/2	1/2	5	364U 365U

(*) Dimension D will never be greater than the above values on rigid mount motors, but it may be less so that shims up to 1/32″ thick (1/16″ on 364U and 365U frames) may be required for coupled or geared machines.

(‡) Dayton motors designated 56H have two sets of 2F mounting holes—3″ and 5″.

(▲) Standard short shaft for direct-drive applications.

(**) Base of Dayton 56HZ frame motors has holes and slots to match NEMA 56, 56H, 143T and 145T mounting dimensions.

(†) Certain NEMA 56Z frame motors have 1/2″ dia. x 1 1/2″ long shaft with 3/64″ flat. These exceptions are noted in this catalog.

(§) Dimension "V" is shaft length available for coupling, pinion or pulley hub—this is a minimum value.

NEMA LETTER DESIGNATIONS FOLLOWING FRAME NUMBER

C—Face mount; see below.
H—Has 2F dimension larger than same frame without H suffix.
J—Face mount to fit jet pumps; see below.
K—Has hub for sump pump mounting.
M—Flange mount for oil burner; 5 1/2″ rabbet dia.
N—Flange mount for oil burner; 6 3/8″ rabbet dia.
T, U—Integral HP motor dimension standards set by NEMA in 1964 and 1953, respectively.
Y—Non-standard mounting; see mfr's drawing.
Z—Non-standard shaft extension (N-W, U dim.)

For their own identification, manufacturers may use a letter before the NEMA frame number. This has no reference to mounting dimensions.

Courtesy of W. W. Grainger, Inc.

INDEX

Abbreviations 333
Air density 344
Air Flow Measuring Stations, estimating 293
Air Handling Units, estimating 277-79
Amp Draws 22, 350

Balancing (see testing and balancing)
Baseboard, estimating 296-97
Bin method 201
Boiler, estimating 294
Brake horsepower 338
Budget estimating controls 108

Centrifugal fans 64
Chillers, estimating 299-300
Coil controls
- chilled water 100
- direct expansion 101
- electric heating 103
- face and bypass
- hot water 103
- steam heating 102

Coils, estimating 298, 305-06
Concepts 4
Condensing units, estimating 304
Controllers 39
Control systems
- central 89
- DDC 88, 91
- types 81

Controls
- budget estimating 108
- building pressurization 97
- coil controls 100
- components 84
- DDC 87, 91
- definitions 88
- designing 173
- economizer system 97, 99
- EMS costs 107
- estimating, 314-17
- function points 93
- morning warm up 105
- night heating 105
- pneumatic 87
- resets 95
- return air tracking 94
- static pressure 83
- sub-systems 81
- symbols 86
- temperatures 83
- types of systems 81
- volumetric 83

Cooling towers, estimating 301-02

Degree Days 194
Designing VAV systems
- air distribution systems 172
- air quantities 175
- control systems 173
- duct design 173

- equipment 172
- hydronic distribution 172
- loads 171
- outlet selection 178
- overall procedure 171
- return plenum ceilings 172
- throttling ratios 177
- type systems 172
- zoning 171

Direct digital controls 87
- components 92
- diagram of system 91

Diversity 9, 177, 208, 262

Duct design
- air quantities 175
- criteria 172
- diversity 177
- return plenum ceilings 177
- throttling ratios 177
- VAV 173

Ductwork
- installed costs 270, 271
- types 266

Economizer system 99

Electrical formulas 351

Energy consumption savings 180

Energy management systems 88
- costs 107
- estimating 318

Energy waste
- minimizing mixed air 95
- minimizing supply air 96

Estimating
- air flow measuring stations 293
- air handling units 277-79
- baseboard 296-97
- boilers 293
- chillers 299-300
- coils 298, 305-06
- condensing units 304
- controls 314-17
- cooling towers 301-02
- copper tubing 273
- ductwork 270, 271
- energy management systems 318
- existing buildings 326-27
- fans 282-86
- heat pumps 302-03
- insulation 307-13
- items to include 263
- piping 270
- procedure 263, 265
- register, grilles, diffusers 288-90
- retrofits 318-21
- roof top units 280-81
- steel piping 272
- testing and balancing 322-26
- valves 274
- VAV terminals 290-92

Existing buildings, estimating 326-27

Fan curves 75

Fan laws 338

Fan, basics
- centrifugal 64
- changing CFM's 78
- classes 77
- curves 75
- drive arrangements 69

estimating 282-86
inlets 69
performance 79
rotations 69
vaneaxial 64
wheels 66
Fans, testing and balancing 214
Fans, varying volume methods
eddy current drives 62
fan bypass 64
inlet vane dampers 56
inverters 57
riding fan curve 53, 55
shrouds 62
vaneaxial fans 59
variable pitch sheaves 61
Flow hoods 234
Formulas
air flow 336
break horsepower 338
coil chiller gpm 348
elec power, pump 347
electrical 351
fan laws 338
heat transfer air 340
total pressure 336
velocity pressure 337
Fuel heating values, costs 192, 193

Grille, Registers, Diffusers
estimating 288-89

Heat losses 183
Heat pumps, estimating 302-03
Heat transfer
latent 340
sensible 340

Horsepower saved 52

Instrument
ammeter 224
anemometers 237
electro manometers 229
flow hoods 234
manometers 227
pitot tubes 233
tachometers 225
testing and balancing 224
velometers 236
Insulation, estimating 307-13

Loads, heating, cooling
bin method 201
cooling load temp
differences 199
degree days 194
factors 7
full load operation hours 198
heat gain, occupants 190
hour by hour 202
instantaneous 183
peak load 7
profiles 8
R and U factors 189
sunload 5
zone loads 5
Leak testing 239

Manometers 227, 228, 229
Morning warmup 105, 261
Motor efficiencies 350
Motors 211

Negative pressure in
buildings 260

Night heating 105, 261

Occupants, heat gain 190
Outlets selection 178
Outside air
economizer 97
fixed 98
Overloads 211

Piping
estimating steel piping 272
estimating copper tubing 273
Pitot tubes 221, 233
Pneumatic controls 87
Points control function 93
Power factors 350
Pressurization of buildings 97
Proportionate balancing 223
Problems, VAV 2
Psychrometric chart 249

R factors 189
Return plenum ceilings 177
Resets
mixed air 95
supply air 95
Retrofits, estimating 318-21
Return air tracking 94
Roof top units, estimating 281

Sensors 37
Static pressure sensors 83
setting 50
Starter sizes 350
Systems, VAV
basic components 15
secondary 13
true 2
types, list 13
types
combined interior and perimeter 18
cooling only interior 16
damper terminal 31
dual duct, dual fan 21
dual duct, single fan 21
fan discharge 30
fan powered 23
multizone 25
pressure classification 14
riding fan curve 29
separate interior and perimeter 19
system powered 27
terminal bypass 30
VAV induction 25
understanding 11

Tachometers 225
Terminals, constant volume 35
Terminals, VAV
calibrating 48
classifications 33
components 37
control power 33
controllers 37
normally open or closed pressure independent or dependent 34
purpose 36
sensors 37
setting S.P. sensors 50

types list 38
Testing and Balancing
basics 205
check VAV terminals 220
coils 215
damper 216
diversity in system 208
estimating 322-26
fan readings 217
fans 214
filters 215
final settings and readings 223
heart of system 210
instruments 224
leak testing 239
motor and starter 211
overall procedure 206
pitot tube traverse 221
preliminaries 210
proportionate balance 223
Trouble shooting air systems 247
building negative pressure 260
checking procedure 250
debugging 257
diversity problem 262
drafts 257
extra energy costs 249
fan problems 254, 255, 256, 260, 261
not enough air 253
poor zoning 257
system imbalance 256
too hot or too cold 252
typical problems
causes 247, 248
Types of VAV systems
bypass 30
cooling only 16
dual duct 21
fan powered 23
induction 25
reheat 18
system powered 27

U factors 189

Valves, estimating 274
Vaneaxial fans 64
Velocity pressure 337, 345, 346
Velometers 236

Weather data 188

Zones, zoning
design 171
trouble shooting 257
zone loads 5